MATHEMATICAL MODELS FOR THE STUDY OF THE RELIABILITY OF SYSTEMS

This is Volume 124 in
MATHEMATICS IN SCIENCE AND ENGINEERING
A Series of Monographs and Textbooks
Edited by RICHARD BELLMAN, *University of Southern California*

The complete listing of books in this series is available from the Publisher upon request.

Mathematical Models for the Study of the Reliability of Systems

A. KAUFMANN

Université de Louvain, Belgium

D. GROUCHKO

R. CRUON

Anciens Élèves de l'Écoles Polytechnique, France

Translated by Technical Translations

ACADEMIC PRESS New York San Francisco London 1977

A Subsidiary of Harcourt Brace Jovanovich, Publishers

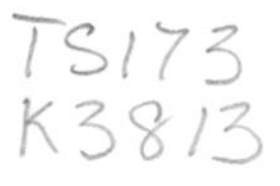

ACADEMIC PRESS, INC.
111 Fifth Avenue, New York, New York 10003

United Kingdom Edition published by
ACADEMIC PRESS, INC. (LONDON) LTD.
24/28 Oval Road, London NW1

Library of Congress Cataloging in Publication Data

Kaufmann, Arnold.
Mathematical models for the study of the reliability of systems.

(Mathematics in science and engineering series ;)
Translation of Modèles mathematiques pour l'étude de la fiabilité des systèmes.
Bibliography: p.
Includes index.
1. Reliability (Engineering)–Mathematical models. I. Grouchko, Daniel, joint author. II. Cruon, R., joint author. III. Title. IV. Series.
TS173.K3813 620'.004'5 76-19489
ISBN 0-12-402370-3

PRINTED IN THE UNITED STATES OF AMERICA

Original edition, Modèles Mathématiques pour l'Étude de la Fiabilité des Systèmes, copyright Masson et Cie, Éditeurs, Paris, 1974.

CONTENTS

PREFACE

The notion of a system is found in all organizations of the living world and of the world created by men. The analysis of complex systems with numerous elements, often interdependent, involves the use of methods that are not yet sufficiently general and that are often too theoretical and of rather academic interest. In spite of this, the analysis and synthesis of complex networks necessarily progresses as much in technological domains as in economical and presently also in biological areas.

The reliability of a complex system, that is, one containing a rather large number of interactive elements, is a question that interests almost all engineers and technicians of all disciplines. One finds that in the past 15 years methods in this domain have been very obviously improved and that it is possible to present concretely a sufficiently global theory for approaching the most frequently encountered cases. When we must study large systems, it is evidently appropriate not to neglect to take into account the aspect of reliability. But if the theories available for the study of the technological or economical aspects of large systems are still insufficiently strong, it appears that in the domain of reliability this is not so. This permits us to present the first published work on a general theory of the reliability of systems. Indeed, this theory is susceptible to new developments; however, this is the case for all theories, which by definition are works in progress. This work may, meanwhile, aid engineers in attacking more efficaciously a great number of difficult problems.

There exist a number of works that treat reliability, but none treats systems completely; many are content to study the reliability of a component in some often deep aspect, with only a few pages devoted to combinations of elements. This book is intended to fill this gap.

In Chapter I we review the now classical notions concerning the lifetime of an element; this is done so that the notation subsequently employed will have been well explicated. Chapter II introduces the very important notion of a survival function with increasing failure rate. In Chapter III the general method for studying systems with n components is developed from the point of view of the logic of their functioning. The notions of *structure function* and *reliability network* are presented through applications. Noncomplementable bivalent variables and two dual operations are used. A reader who has an appropriate mathematical background will be pleased to note that the entire theory considered is in fact that of free distributive lattices with n generators. Chapter IV presents the application to the study of the reliability of systems. The Moore–Shannon theorem plays a central role. All this leads to the notion of redundance, which is most important for engineering applications; Chapter V is devoted to this topic. Finally, Chapter VI treats the case of dual failures.

Much remains to be written on the subject, for example, on economic aspects, cannibalization, replacement, and maintenance of systems. However, it is hoped that this volume will be useful to those who have the responsibility of constructing and maintaining complex technological structures.

LIST OF SYMBOLS[1]

A_k	number of links having k components (see Theorem 17.II)
a	link of a structure (see (16.3))
B_k	number of cuts having k components (see Theorem 17.II)
b	cut of a structure (see (16.10))
$E[T^k]$	noncentral moment of order k of the random variable T
e	set of components of a system
e_i	component of a system
G	graph (see Section 18)
$h(p)$	reliability function of a system (see (25.7))
$i(t)$	probability density of lifetime (see (4.2))
$L(t) = \Lambda(t)/t$	see (12.1)
ln	natural logarithm
log	base-10 logarithm
$\mathbf{N} = \{0, 1, 2, \ldots\}$	set of nonnegative integers
O	origin of a reliability network (see Section 19)
p	vector of component reliabilities (see Section 25)
p_i	reliability of a component (probability that it will function at a certain instant t)
$p(n)$	probability that lifetime will equal n (law of type I; see p. 13)
$p_c(n)$	conditional probability of failure (law of type I; see (4.11))
R	set of real numbers
$\mathbf{R}^+$	set of nonnegative real numbers
$\mathbf{R}_0{}^+$	set of positive real numbers
$\mathscr{R}$	reliability network
$\bar{\mathscr{R}}$	reliability network dual to network $\mathscr{R}$
r	number of components or order of a system
S	set of vertices of a graph (see Section 18)
S	system
$S(X)$	random number of components in a good state
T	*lifetime*
$\bar{T} = E(T)$	mathematical expectation of lifetime

[1] Sets are designated with boldface letters: **a**, **A**.

U	set of arcs of a graph (see Section 18)
u_i	arc of a graph
$v(t)$	survival function (complementary distribution function lifetime): $v(t) = \text{pr}\{T > t\}$
$v_a(t)$	survival function of equipment with initial age a
$v_g(t; a)$	survival function of equipment guaranteed until age a
$w(t)$	see (5.7)
X_i	(random) state of component e_i (see (25.2))
x	state of the set of components (see Section 15)
x_i	state of component e_i (see Section 15)
Z	terminal point of a reliability network (see Section 19)
$\binom{n}{k}$	number of combinations of n objects taken k at a time. For n negative, see (6.74)
Δ	mapping associating a component of the system with each arc of a graph
$\Lambda(t)$	cumulative failure rate (see (4.18) and (4.20))
λ	length of a system (see p. 66)
$\lambda(t)$	instantaneous failure rate
μ	width of a system (see p. 66); path of a graph
$\mu(t; x)$	failure rate in an interval $]t, t + x]$, open on the left and closed on the right (see (10.2))
σ_T^2	variance of the random variable T
$\Phi(X)$	random structure function (see (25.5))
$\Phi(t)$	distribution function of lifetime: $\Phi(t) = \text{pr}\{T \leq t\}$
$\varphi(x)$	structure function of a system (see Section 15)
$\bar{\varphi}(x)$	structure function dual to $\varphi(x)$
$\varphi_s(x)$	structure function put in simple form (see p. 95)
$\succcurlyeq, \preccurlyeq$	nonstrict order relation between r-tuples (see (15.15) and (15.16))
$\succ, \prec$	strict order relation between r-tuples (see (15.17) and (15.18))
$\simeq$	equivalence relation between networks and structure functions (see Section 20)
$[x]$	largest integer less than or equal to x
$[a, b]$	interval (a, b) closed on the left and on the right
$\in$	membership relation of an element to a set
$\subset$	nonstrict inclusion relation of one set in another
$\subset\subset$	strict inclusion relation of one set in another

CHAPTER I

LIFETIME OF A COMPONENT

1 Introduction

The simplest way of studying a component, from the point of view of its reliability and maintenance, is to consider at a given instant whether it is in a good state—functioning or capable of functioning—or in fact has broken down and is not able to provide any service.

This is applicable, for example, to an electric light bulb. In this same case, however, if one looks more closely, one notices that a bulb shows a reduction in its light output as it ages; this diminution generally manifests itself in a detectable fashion (at least with the aid of measuring apparatus) well before the filament rupture that produces the characteristic failure. One will agree, however, that in most applications this reduction in output may be neglected.

This will not necessarily hold true for a piece of complex electronic equipment, a radio receiver for example. This may function in a more or less satisfactory fashion, particularly because of "drift" in the electrical characteristics of its components. Electrical engineers usually distinguish between a "catastrophic" breakdown (a broken circuit or a short circuit, for example), which occurs in an unexpected fashion and has grave consequences, and a failure "through drift," manifesting itself in a progressive manner through changes in the characteristics of the equipment. The wear of mechanical elements presents analogies with this drift of electronic components.

In these cases where the characteristics of the equipment are slowly degraded, whether this occurs in a continuous or erratic fashion, one may fix "tolerance limits" that permit at each instant the unambiguous determination of whether or not the equipment is to be considered as being in a good state. In spite of the arbitrary character of these tolerance limits, they may be used to reduce the question to the case of equipment having only two possible states.

The problem is further complicated by the diversity of functions that may be performed by complex equipment. To take an extreme example, the failure of the cigarette lighter in an automobile in no way affects driving; its only consequence is to render the auto unable to serve one of its secondary functions, that of serving as a lighter. The decision to regard a component as being or not being in a good state will thus depend on the precise use to which it is to be put.

We shall see later how, by decomposing complex equipment into more simple elements, one may in part surmount the difficulties just mentioned. For the present we shall hypothesize that a component has only two possible states: functioning well or broken down. The considerations that we shall develop in this chapter will be directly applicable to certain relatively simple equipment, and will provide a point of departure for the study of more complex systems.

2 Age and Lifetime of a Component

In addition to the hypothesis discussed above, according to which a component has only two possible states, functioning or failure, we shall suppose that failure is irreversible. We thus discard the possibility of an intermittent failure, capable of disappearing without external intervention. Moreover, we shall not preoccupy ourselves here with the possibilities of repair of the component. The "life" of a component then follows a very simple scheme: the new component is put into service, functions for a certain time, then "dies."

If one has been able to observe the life of a great number of components, the classical methods of descriptive statistics permit the presentation of the results of observations in a simple fashion. In order to fix the ideas, suppose that 1000 components have been put into service at the date $t = 0$.[1]

At each date t we shall determine the number of components that have become disfunctional in the interval $]t, t + 1]$, $t = 0, 1, 2, 3, \ldots$. We suppose

[1] In fact, it is not necessary that the parts be put into service simultaneously provided that for each of them service time is calculated with respect to its arrival in service. Difficulties may arise, however, if the conditions of functioning evolve over the course of time since then they may not be the same for all the components, but may vary as a function of the initial service date.

that a faulty component is never replaced. The result of the observations may be represented in a histogram on which the relative frequencies are shown (Fig. 2.1a). We may then trace a cumulative histogram (Fig. 2.1b) which indicates at each date t the proportion of components out of service in the interval $]0, t]$.

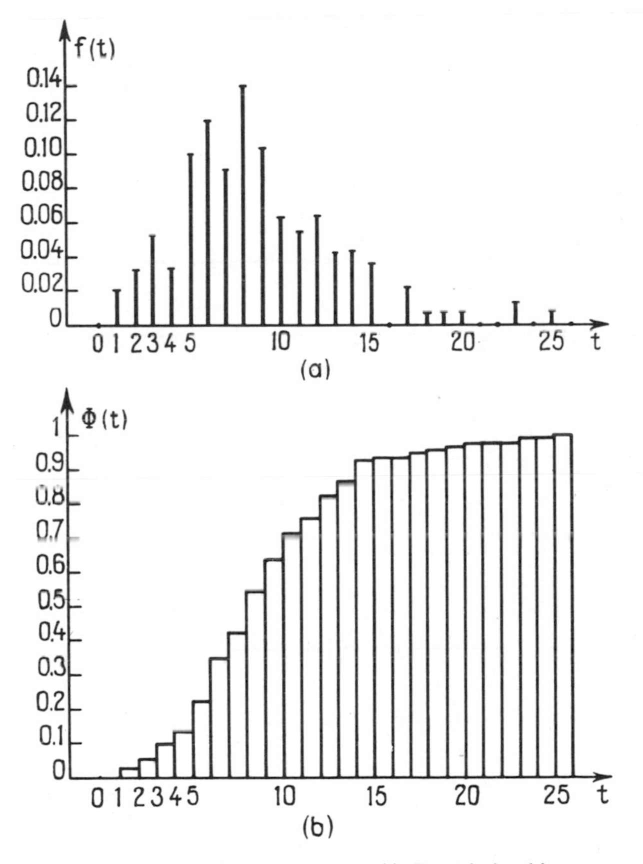

FIG. 2.1. (a) Relative histogram. (b) Cumulative histogram.

A complementary diagram may be easily obtained from that of Fig. 2.1b by evaluating, from the cumulative relative frequencies $\Phi(t)$, the function $v(t) = 1 - \Phi(t)$. Such a diagram (Fig. 2.2) permits evaluation, at date t, of the number of components that are still in service. The diagram of Fig. 2.1b constitutes a "mortality statistic"; the complementary diagram (Fig. 2.2) is a "survival statistic."

Presented in the form of a continuous curve (Fig. 2.3), the histogram of Fig. 2.2 then takes the name "survival curve."

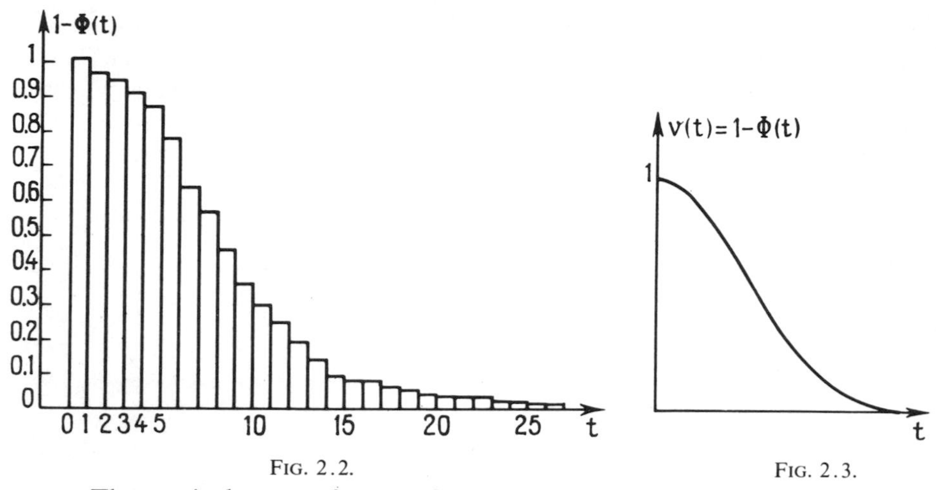

FIG. 2.2. FIG. 2.3.

The survival curve of a set of eternal components will have the shape given in Fig. 2.4a; that of a set of rigorously identical components used under equally rigorously identical conditions will have the form given in Fig. 2.4b. Before some date θ all of the components will have been in service, and after this date none of them will be. The curves given in Figs. 2.4c and 2.4d correspond to more realistic hypotheses.

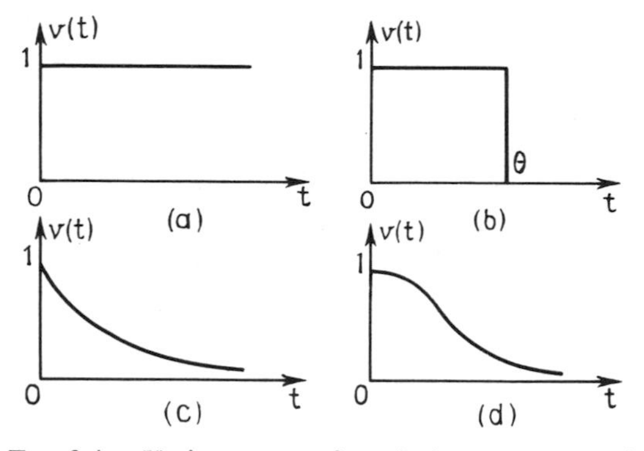

FIG. 2.4. Various types of survival curves (see text).

Choice of a Parameter Measuring Age. The type of statistical description mentioned above raises no difficulty in principle when the equipment functions continually from the time it is put into service until its death. With the reservation that the conditions for functioning be sufficiently well defined, the number of hours (or any other convenient unit of time) of functioning indeed measures, at each instant, the "amount of use" that the equipment has provided until the instant being considered.

If equipment is used in an intermittent fashion, the simplest solution is evidently to measure the age of the component at a given instant by the sum of the effective durations of use. This is valid, however, only if the component does not deteriorate when it is not used (as the advertisements of a brand of batteries claim) and if starting and stopping have no detrimental influence on lifetime. Consider again the example of an electric light bulb already used in Section 1. It is clear that the first of the two conditions above is satisfied in this case; but it is possible that the thermal shocks sustained by the filament during lighting and extinction have a nonnegligible influence. In order to determine this, it would be necessary to compare the lifetimes (computed as the total number of hours of functioning) for bulbs used continually and for bulbs sustaining a large number of illuminations and extinctions. If, in the second case, one obtains a shorter mean lifetime, one may propose measuring the age of a bulb with a quantity of the form $h + kn$, h being the number of hours of functioning and n the number of times the bulb was illuminated. The coefficient k will of course be chosen in such a fashion that the lifetime of a bulb will be statistically the same regardless of mode of use.

In practice, the lack of statistical data very often leads to the use of a very simple solution. The age of a component is thus measured by the number of hours of functioning (the present case), by the number of kilometers traveled by a vehicle, or even by the number of uses: the number of openings (or of closures) for a relay, the number of landings for the tire of an airplane, etc.

We are beginning, however, to measure (particularly in the case of electronic materials) and to characterize in less gross fashion the age of a component. For example, one generally estimates that the rate of deterioration of an electronic component that is not under tension is of the order of $\frac{1}{30}$ of the rate of deterioration while functioning. One may thus express the age of a component by $t_F + t_S/30$, where t_F designates the number of hours of functioning and t_S the number of hours at rest or in storage. Unfortunately, the influence of the number of times that the component is put into use is still not known.

Influence of Environmental Conditions. We have just seen, in the case of the alternation of periods of functioning and rest for a component, a first example of the influence of conditions of use of a component on its lifetime. The environment in which the apparatus is placed likewise has considerable importance.

It would readily be conceded that equipment used in a laboratory, thus in a calm atmosphere and by people accustomed to handling delicate apparatus, is not at all subject to the same constraints as is equipment used in a work yard or put aboard a vehicle. In the second case mechanical vibrations, thermal shocks, humidity, and shortcomings of the operators all may considerably shorten the lifetime of the equipment. For example, for evaluation of the reliability of electronic equipment mounted on airplanes, one is led to introduce a "K factor" that multiplies the rate of deterioration established for the equipment on the ground, and whose value may range from about ten to a thousand.

Similarly, for electronic components, it is necessary to be able to take into account, in addition to the general environmental factors of the equipment, particular conditions of use due to their places in the circuit. For example, a transistor used at $\frac{1}{10}$ or $\frac{1}{100}$ of its nominal power will clearly have a longer lifetime than if it were used at the limits of its possibilities.

We refer the reader to the specialized literature[2] for more details. It is necessary, however, to remember that *the notion of the lifetime of a component only has sense when the conditions of use have been precisely defined.*

We shall suppose in the remainder of this work that the equipment considered functions in well-defined conditions, and that one has chosen a parameter that well represents the "amount of use" that was in question above. This parameter will be called the *age* of the equipment. The *lifetime* of equipment will then be the age that it has attained at its "death," that is, when it has fallen in failure.

3 Survival Function

The discussion of Section 2 shows that the lifetime of equipment may not be described in a precise fashion without the language of the theory of probability. We thus suppose that the lifetime of a component may be represented by a random variable T, for which the *survival function* $v(t)$ is defined by

$$v(t) = \text{pr}\{T > t\}. \tag{3.1}$$

This fundamental hypothesis determines all the theory that will be developed in this work. The practical application of this theory evidently runs into a sizeable problem—that of the estimation of the probability law of T. In this work we do not attack this statistical aspect of the study of the reliability of equipment, for which we refer the reader to works on statistics. Our goal is only to propose abstract *models*, allowing the description in rigorous language of the principal phenomena related to the reliability of

[2] See, for example, for electronic components and for certain electric and electromechanical components, the collection of reliability data of the Centre National d'Etude des Telecommunications, or various American documents ("R.A.D.C. Notebook," "Handbook MIL HDBK" 217 A or B, etc.).

equipment. We believe that even in cases where the given data are not sufficiently abundant to allow complete use of these models, this will permit at least some understanding of the phenomena being studied. On the other hand, grasping an awareness of the possibilities that follow from a quantitative analysis of problems solved by the management of equipment will lead, we hope, those responsible for this management to gather systematically the necessary information.

If one accepts the hypothesis above, the probability that a component has a failure at a date T earlier than t is

$$\text{pr}\{ T \leqslant t \} = 1 - v(t) = \Phi(t) . \tag{3.2}$$

According to the usual terminology of the calculus of probability, $\Phi(t)$ is a distribution function and $v(t)$ the complementary distribution function. In problems of equipment failure, this last function in current usage is called a "survival function," as we have previously indicated.

Following the choice of a parameter measuring the age of a component (Section 2), the random variable T will take its values in an interval of the real numbers $\mathbf{R}$, most often in the interval $[0, \infty)$, or in a denumerable set, most often $\mathbf{N} = \{ 0, 1, 2, \dots \}$. The convenience of the analysis likewise plays an important role in the choice of the set of possible values for the random variable T. For example, the lifetime of an electric relay is expressed as an integer (the number of openings or closings of the contacts); but in such a case the numbers with which one will be concerned will always be very large, and it may be convenient to consider T as being able to vary continuously.

Three types of probability laws will be used[3]:

TYPE *I*. The random variable T representing the lifetime of a component is a discrete variable. The survival function is a "step function."

TYPE *IIa*. The random variable T takes its values in $[0, \infty)$; its distribution function $\Phi(t)$, and as a consequence its complementary distribution function $v(t)$, are continuous for all t. Also, $\Phi(t)$ and $v(t)$ admit a derivative in the interval where they are defined.

TYPE *IIb*. The random variable T takes its values in $[0, \infty)$, where $\Phi(t)$ and $v(t)$ are continuous piecewise and make at least one jump.

A distribution function of a random variable of type IIb may be considered as formed by the superposition of two functions, one a step function

[3] In various works type IIa is called type II, whereas type IIb is referred to as of mixed type. The indicated classification is not exhaustive, but it includes all functions that can be conveniently used in practice.

and the other continuous, these functions being mixed by a given convex weighting. Figure 3.1 gives examples of survival functions corresponding to each of these three types.

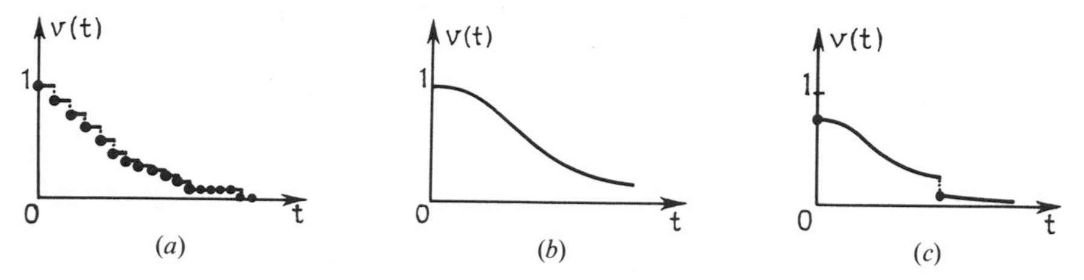

FIG. 3.1. (a) Example of a law of type I; (b) example of a law of type IIa; (c) example of a law of type IIb.

Concerning the interval of definition of the survival function, we may always take it to be equal to $(-\infty, \infty)$; the condition $T \geqslant 0$ and (3.1) imply that $v(t) = 1$ for $t < 0$ for types IIa or IIb. Concerning the case of a law of type I, we most often use the set $\mathbf{N} = \{0, 1, 2, 3, \ldots\}$ where each value of $t \in \mathbf{N}$ will be called a "date."

Through abuse of language we shall often write "survival law" or "mortality law" for "probability law of survival" or "probability law of mortality."

In order that a given function $v(t)$ be a complementary distribution function of a random variable $T \geqslant 0$, it is necessary that:

(3.3)
(1) $v(t) = 1$ for $t < 0$,
(2) $v(\infty) = 0$, and
(3) $v(t)$ is monotone nonincreasing on $(-\infty, \infty)$.

It follows from definitions (3.1) and (3.2), which we adopted for the distribution function and the complementary distribution function, that the

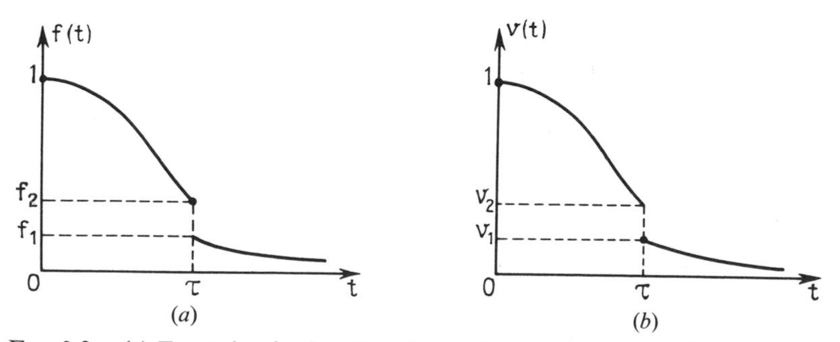

FIG. 3.2. (a) Example of a function $f(t)$ continuous on the left. One has $f(\tau) = f_2$. (b) Example of a function $v(t)$ continuous on the right. One has $v(\tau) = v_1$. We have chosen the convention of part (b).

functions $v(t)$ being considered in the present work will always be continuous and differentiable on the right. Figure 3.2 shows the consequences of such a choice.

The above details will take on importance when using functional transformations of Laplace type (operator, operational, or symbolic calculus), or for increasing and decreasing failure rate functions (Section 10).

4 Failure Probability. Failure Rate

Consider a survival function $v(t)$ of type IIa; we shall then define a "probability density of lifetime" $i(t)$ such that

$$(4.1) \qquad i(t).\mathrm{d}t = \mathrm{pr}\,\{\, t < T \leqslant t + \mathrm{d}t \,\} = \mathrm{pr}\,\{\, T \leqslant t + \mathrm{d}t \,\} - \mathrm{pr}\,\{\, T \leqslant t \,\},$$

thus

$$(4.2) \qquad i(t) = \frac{\mathrm{d}\Phi}{\mathrm{d}t} = -\frac{\mathrm{d}v}{\mathrm{d}t} \quad \text{or} \quad v(t) = \int_t^\infty i(u)\,\mathrm{d}u\,.$$

Quantity (4.1) represents the probability that a component has a failure in the interval $]t, t + \mathrm{d}t]$. Figure 4.1 shows two characteristic aspects of the curve $i(t)$: exponential type and bell shaped, corresponding to the survival curves of Figs. 2.4c and 2.4d.

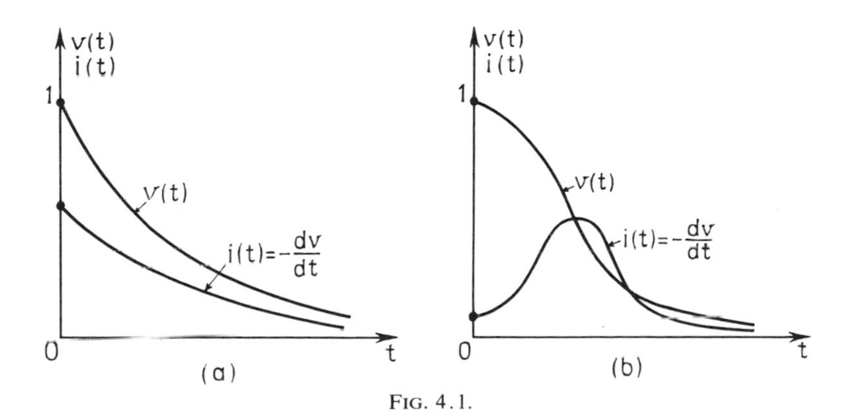

FIG. 4.1.

When we have a law of type I, t takes its values in $\mathbf{N} = \{\, 0, 1, 2, 3, \ldots \,\}$, and the probability that the component has a failure at date $n + 1$ is

(4.3) $$\operatorname{pr}\{ T = n + 1 \} = \operatorname{pr}\{ T \leqslant n + 1 \} - \operatorname{pr}\{ T \leqslant n \} = \Phi(n + 1) - \Phi(n), \qquad n = 0, 1, 2, \ldots$$

with

$$\operatorname{pr}\{ T = 0 \} = \Phi(0);$$

or what is the same

(4.4) $$\begin{aligned}\operatorname{pr}\{ T = n + 1 \} &= (1 - \operatorname{pr}\{ T > n + 1 \}) - (1 - \operatorname{pr}\{ T > n \}) \\ &= \operatorname{pr}\{ T > n \} - \operatorname{pr}\{ T > n + 1 \} \\ &= v(n) - v(n + 1), \qquad n = 1, 2, 3, \ldots\end{aligned}$$

with

$$\operatorname{pr}\{ T = 0 \} = 1 - v(0).$$

We then have

(4.5) $$\begin{aligned}\Phi(n) &= \operatorname{pr}\{ T=0 \} + \operatorname{pr}\{ T=1 \} + \cdots + \operatorname{pr}\{ T=n-1 \} + \operatorname{pr}\{ T=n \} \\ &= \operatorname{pr}\{ T \leqslant n \}\end{aligned}$$

and

(4.6) $$\begin{aligned}v(n) &= \operatorname{pr}\{ T = n + 1 \} + \operatorname{pr}\{ T = n + 2 \} + \cdots \\ &= \operatorname{pr}\{ T > n \}.\end{aligned}$$

Returning to the case of a law of type IIa, we may write

(4.7) $$\operatorname{pr}\{ t < T \leqslant t + \mathrm{d}t \} = \operatorname{pr}\{ T > t \} . \operatorname{pr}\{ T \leqslant t + \mathrm{d}t \mid T > t \}$$

or, further, if $v(t) = \operatorname{pr}\{ T > t \} > 0$,

(4.8) $$\begin{aligned}\operatorname{pr}\{ T \leqslant t + \mathrm{d}t \mid T > t \} &= \frac{\operatorname{pr}\{ t < T \leqslant t + \mathrm{d}t \}}{\operatorname{pr}\{ T > t \}} \\ &= \frac{i(t)}{v(t)} . \mathrm{d}t \\ &= -\frac{v'(t)}{v(t)} . \mathrm{d}t \quad \text{where} \quad v'(t) = \frac{\mathrm{d}v}{\mathrm{d}t} .\end{aligned}$$

We designate expression (4.8) by $\lambda(t)$:

(4.9) $$\lambda(t) = -\frac{v'(t)}{v(t)} .$$

This function is called the "instantaneous failure rate" or more simply the "failurc rate." The quantity $\lambda(t)\,\mathrm{d}t$ represents the probability that the component fails in the interval $]t, t + \mathrm{d}t]$, *knowing that it is in a good state at the instant t*. This is thus a conditional probability, whereas $i(t)\,dt$ is the a priori probability that the component fails in $]t, t + dt]$. Statistically, $\lambda(t)\,\Delta t$ may be estimated by the ratio of the number of components that have failed between t and $t + \Delta t$ to the number of components in a good state at the instant t.

If we consider a law of type I instead of type II, we will have

$$\text{pr}\{T = n + 1\} = \text{pr}\{T > n\}\, p_c(n) \tag{4.10}$$

where $p_c(n) = \text{pr}\{T = n + 1 \mid T > n\}$. From (4.10) we may deduce

$$p_c(n) = \frac{\text{pr}\{T = n + 1\}}{\text{pr}\{T > n\}} = \frac{v(n) - v(n + 1)}{v(n)}. \tag{4.11}$$

The quantity $p_c(n)$ will be called the "conditional failure probability." Note that

$$p_c(-1) = \frac{1 - v(0)}{1} = \text{pr}\{T = 0\}. \tag{4.11'}$$

We often use other names for the failure rate: "hazard rate," "mortality rate," or "strength of mortality" (an actuarial usage); "Mills index" is used if one is considering the normal law.

Survival Law Defined by a Rate of Deterioration. For a law of type IIa, Eq. (4.9) may be written as

$$\frac{dv}{dt} + \lambda(t).v(t) = 0. \tag{4.12}$$

We also have

$$v(0) = 1$$

since $v(t)$ is by hypothesis a law of type IIa.

If we suppose that $\lambda(t)$ is given, Eq. (4.12) is a differential equation in $v(t)$. Its solution is

$$v(t) = \exp\left(-\int_0^t \lambda(u)\, du\right). \tag{4.13}$$

We now consider two particularly interesting cases:

$$1)\ \lambda(t) = \lambda_0, \qquad t \geqslant 0, \qquad \lambda_0 > 0. \tag{4.14}$$

We then have

(4.15) $$v(t) = e^{-\lambda_0 t}.$$

(4.16) $$2)\ \lambda(t) = k_0 t, \qquad t \geqslant 0, \qquad k_0 > 0.$$

We then have

(4.17) $$v(t) = e^{-k_0 t^2/2}.$$

Thus if $\lambda(t)$ is constant, the survival law is an exponential law; and it is easy to see that the converse is equally true. If $\lambda(t)$ is a linearly increasing function, the curve $v(t)$ is a bell curve truncated at the origin.

We take note of several mathematical properties:

(a) The function $\lambda(t)$ is always nonnegative (evident).
(b) The function $\lambda(t)$ is not necessarily monotone.
(c) One may have $\lambda(0) > 0$.

In practice, one typically finds the case where the failure rate $\lambda(t)$ is increasing. There are cases, however, where $\lambda(t)$ initially decreases; one may also show practically that $\lambda(0)$ is generally nonzero.

In Sections 10 and 11 we shall study in particular two general types of survival curves:

(1) survival curves for increasing failure rates (IFR), and
(2) survival curves for decreasing failure rates (DFR)

without excluding the case where the rate is constant (exponential survival curve). Of course, one may be interested from a theoretical point of view in other cases less important in practice.

Cumulative Failure Rate. Logarithmic Survival Function. In certain applications, we will be interested in the function

(4.18) $$\Lambda(t) = \int_0^t \lambda(u)\,du,$$

which is called the "cumulative failure rate" or "logarithmic survival function."

According to (4.13) we have

(4.19) $$v(t) = e^{-\Lambda(t)}$$

or

(4.20) $$\Lambda(t) = -\ln v(t),$$

the notation "ln" signifying the natural logarithm.

The condition $v(\infty) = 0$ implies that the function $\Lambda(t)$ tends toward $+\infty$ when t tends toward $+\infty$; the condition $v(t) \leqslant 1$ implies that $\Lambda(t) \geqslant 0$.

Relation (4.20) permits the extension of the definition of cumulative failure rate to laws of type IIb, and thus to laws of type I. In the latter case, the cumulative failure rate presents, however, less interest than that for laws of type II.

5 Moments of a Survival Law. Mean Failure Age

Let $\Phi(t)$ be the distribution function of the lifetime T. The moments of this law are

(5.1) $$E[T^k] = \int_0^\infty t^k \, \mathrm{d}\Phi(t) .$$

By considering the survival function

(5.2) $$v(t) = 1 - \Phi(t) ,$$

Eq. (5.1) may also be written as

(5.3) $$E[T^k] = -\int_0^\infty t^k \, \mathrm{d}v(t) .$$

In the case of a law of type IIa we may write

(5.4) $$E[T^k] = \int_0^\infty t^k . i(t) \, \mathrm{d}t$$

where $i(t)$ is the probability density of the lifetime defined by (4.2).

In the case of a law of type I where t takes its values in $\mathbf{N} = \{0, 1, 2, 3, \ldots\}$, relation (5.1) becomes

(5.5) $$E[T^k] = \sum_{n=0}^{\infty} n^k . p(n)$$

where $p(n) = \mathrm{pr}\,\{ T = n \}$.

Calculation of the moments may be carried out using characteristic functions (or generators, for (5.5)), or also by using convenient functional transformations (Laplace or Carson–Laplace, z transform, etc.). On the subject of general methods for the calculation of moments, the reader is referred to a course on the theory of probability.[4] We proceed, for now, to concern ourselves with the calculation of the mean and variance of T.

[4] See, for example, A. Kaufmann, *Cours Moderne de Calcul des Probabilités*. Albin-Michel, Paris, 1965.

Descamps [15] has given a convenient method for calculating

$$\overline{T} = E[T] \qquad \text{and} \qquad \sigma_T^2 = E[(T - \overline{T})^2]\,,$$

in the case, which interests us here, of a nonnegative random variable T.

Suppose that there exists an $\alpha > 0$ such that

$$\lim_{t\to\infty} [\mathrm{e}^{\alpha t}\, v(t)] = 0\,; \tag{5.6}$$

in other words, $v(t)$ tends exponentially toward 0. Then the function

$$w(t) = \int_t^\infty v(u)\, \mathrm{d}u \tag{5.7}$$

is defined[5] for all $t \geqslant 0$; in addition we have[6]

$$\lim_{t\to\infty} [t^k\, v(t)] = 0\,, \qquad k = 0, 1, 2, \ldots\,, \tag{5.8}$$

$$\lim_{t\to\infty} [tw(t)] = 0\,. \tag{5.9}$$

An interpretation of this function $w(t)$ will be given in Section 8.

We now may write

$$\begin{aligned} \overline{T} = E[T] &= -\int_0^\infty t\, \mathrm{d}v(t) \\ &= -\,[tv(t)]_0^\infty + \int_0^\infty v(t)\, \mathrm{d}t\,. \end{aligned} \tag{5.10}$$

[5] In fact, (5.6) shows that, for any given $\varepsilon > 0$, there exists $\hat{t}$ such that

$$t > \hat{t} \quad \Rightarrow \quad \mathrm{e}^{\alpha t}.v(t) < \varepsilon\,.$$

If one takes $\varepsilon = 1$, one thus has

$$t > t \quad \Rightarrow \quad v(t) < \mathrm{e}^{-\alpha t}\,.$$

One then has

$$w(t) \leqslant w(0) = \int_0^\infty v(t)\, \mathrm{d}t = \int_0^{\hat{t}} v(t)\, \mathrm{d}t + \int_{\hat{t}}^\infty v(t)\, \mathrm{d}t\,.$$

The second term of this last equation is bounded by

$$\int_{\hat{t}}^\infty \mathrm{e}^{-\alpha t}\, \mathrm{d}t = \frac{\mathrm{e}^{-\alpha \hat{t}}}{\alpha} < \infty\,.$$

[6] According to the preceding footnote, $v(t)$ and $w(t)$ are majorized for t sufficiently large, by a function of the form $k\mathrm{e}^{-\alpha t}$; relations (5.8) and (5.9) then follow from a classical theorem.

Thus, according to (5.8),

(5.11)
$$\boxed{\bar{T} = \int_0^\infty v(t)\,\mathrm{d}t = w(0)}\,.$$

On the other hand,

(5.12)
$$E[T^2] = -\int_0^\infty t^2\,\mathrm{d}v(t)$$
$$= -\,[t^2.v(t)]_0^\infty + 2\int_0^\infty tv(t)\,\mathrm{d}t\,,$$

thus from (5.8)

(5.13)
$$E[T^2] = 2\int_0^\infty tv(t)\,\mathrm{d}t\,.$$

By noting that

(5.14)
$$\frac{\mathrm{d}w}{\mathrm{d}t} = -\,v(t)\,,$$

and by carrying out another integration by parts, we develop

(5.15)
$$E[T^2] = 2\int_0^\infty tv(t)\,\mathrm{d}t$$
$$= -\,2[tw(t)]_0^\infty + 2\int_0^\infty w(t)\,\mathrm{d}t\,.$$

Taking (5.9) into account,

(5.16)
$$\boxed{E[T^2] = 2\int_0^\infty w(t)\,\mathrm{d}t}\,.$$

Thus $E[T^2]$ is twice the shaded area below the curve $w(t)$ in Fig. 5.1.

Putting

(5.17)
$$k_1 = \int_0^\infty v(t)\,\mathrm{d}t = w(0)$$

and

(5.18)
$$k_2 = \frac{1}{2}E[T^2] = \int_0^\infty w(t)\,\mathrm{d}t\,,$$

we have finally

$$\overline{T} = k_1 \tag{5.19}$$

and

$$\begin{aligned}\sigma_T^2 &= E[T^2] - (\overline{T})^2 \\ &= 2\,k_2 - k_1^2\,.\end{aligned} \tag{5.20}$$

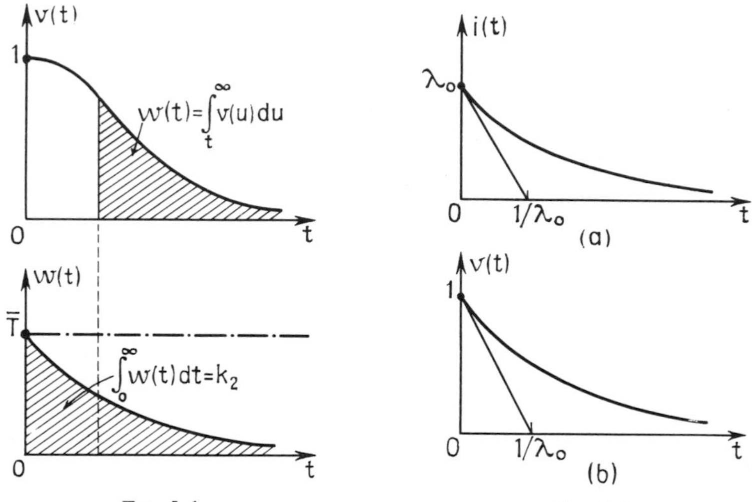

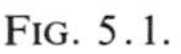
FIG. 5.1.

FIG. 6.1.

6 Principal Survival Laws Used in the Management of Equipment[7]

Laws of Type II

(1) EXPONENTIAL LAW (Fig. 6.1)

$$i(t) = \lambda_0\, e^{-\lambda_0 t}\,, \qquad t \geqslant 0\,, \qquad \lambda_0 > 0\,, \tag{6.1}$$

$$v(t) = e^{-\lambda_0 t}\,, \qquad t \geqslant 0\,, \tag{6.2}$$

$$\lambda(t) = \lambda_0\,, \qquad t \geqslant 0\,, \tag{6.3}$$

$$\overline{T} = 1/\lambda_0\,, \tag{6.4}$$

$$\sigma_T^2 = 1/\lambda_0^2\,. \tag{6.5}$$

(2) GAMMA LAW (Fig. 6.2)

$$i(t) = \frac{\lambda_0\, e^{-\lambda_0 t}(\lambda_0\, t)^{(k-1)}}{\Gamma(k)}\,, \qquad t \geqslant 0,\ k \in \mathbf{R}_0^+\,,\ \lambda_0 > 0 \tag{6.6}$$

[7] See also the article by Morice [41].

where $\Gamma(k)$ is the Eulerian function of the second kind

$$\Gamma(k) = \int_0^\infty x^{k-1} \mathrm{e}^{-x} \, \mathrm{d}x \, . \tag{6.7}$$

We have

$$v(t) = \frac{\lambda_0^k}{\Gamma(k)} \int_t^\infty u^{(k-1)} \mathrm{e}^{-\lambda_0 u} \, \mathrm{d}u \, , \tag{6.8}$$

$$\lambda(t) = \frac{t^{(k-1)} \mathrm{e}^{-\lambda_0 t}}{\displaystyle\int_t^\infty u^{(k-1)} \mathrm{e}^{-\lambda_0 u} \, \mathrm{d}u} \, . \tag{6.9}$$

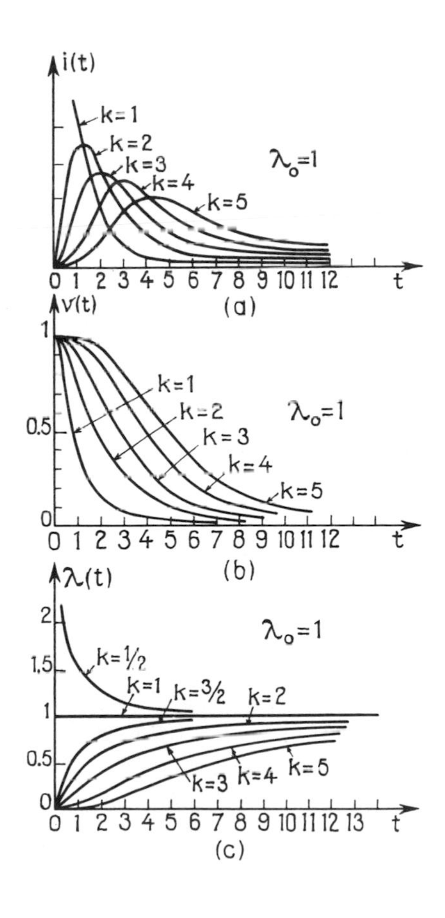

Fig. 6.2.

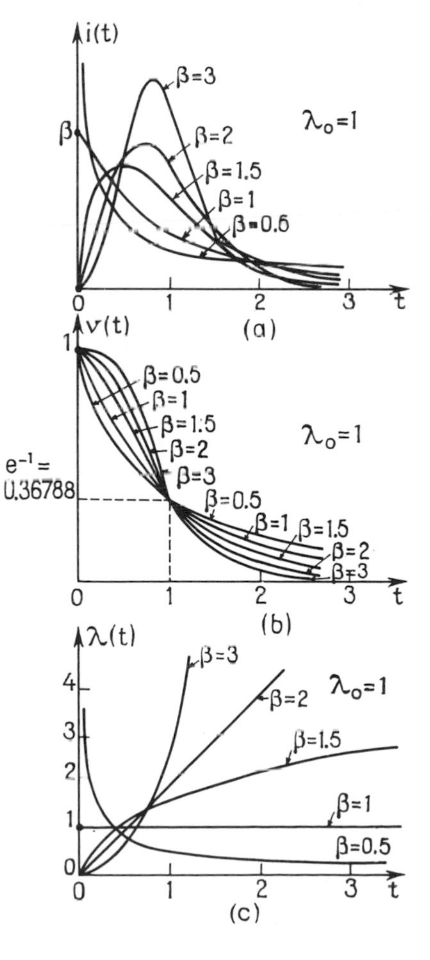

Fig. 6.3.

Figures 6.2a–c give the shapes of the curves $i(t)$, $v(t)$, and $\lambda(t)$, respectively, for various values of k.

On the other hand,

(6.10) $$\overline{T} = \frac{k}{\lambda_0},$$

(6.11) $$\sigma_T^2 = \frac{k}{\lambda_0^2}.$$

When k is an integer greater than or equal to 1, we know that

(6.12) $$\Gamma(k) = (k-1)\,!;$$

the gamma law then also often bears the name "Erlang-k law." This is the probability law of the variable T defined by

(6.13) $$T = T_1 + T_2 + \cdots + T_k,$$

where the k independent random variables T_i all have the same exponential probability law

(6.14) $$i(t) = \lambda_0\, e^{-\lambda_0 t}, \qquad t \geqslant 0,\ \lambda_0 > 0.$$

The survival function then may be written as

(6.15) $$v(t) = e^{-\lambda_0 t} \sum_{r=0}^{k-1} \frac{(\lambda_0\, t)^r}{r\,!}$$

and the failure rate becomes

(6.16) $$\lambda(t) = \frac{\lambda_0^k\, t^{(k-1)}}{(k-1)\,! \displaystyle\sum_{r=0}^{k-1} \frac{(\lambda_0\, t)^r}{r\,!}}.$$

(3) WEIBULL LAW[8] (Fig. 6.3)

(6.17) $$i(t) = \beta\lambda_0(\lambda_0\, t)^{(\beta-1)}\, e^{-(\lambda_0 t)^\beta}, \qquad t \geqslant 0, \quad \beta, \lambda_0 \in \mathbf{R}_0^+,$$

(6.18) $$v(t) = e^{-(\lambda_0 t)^\beta},$$

(6.19) $$\lambda(t) = \beta\lambda_0(\lambda_0\, t)^{(\beta-1)},$$

(6.20) $$\Lambda(t) = (\lambda_0\, t)^\beta.$$

The interest of this law derives from the fact that the failure rate may be, depending on the value of β, increasing, decreasing, or constant (Fig. 6.3c). For $\beta = 1$, we recover the exponential law. For $\beta > 1$, the probability density is represented by a bell curve (Fig. 6.3a). The survival curve is represented in Fig. 6.3b; for $\beta \leqslant 1$, the shape approaches that of the exponential, and for $\beta > 1$, the result is a truncated bell curve.

[8] The reader is referred to the works of Weibull [55,56].

The moment of order r of the Weibull law may be put in the form

$$m_r = E[T^r] = \frac{1}{\lambda_0^r}\Gamma\left(\frac{r}{\beta} + 1\right), \tag{6.21}$$

where $\Gamma(x)$ is as before the Eulerian function (6.7). The moments of orders 1 and 2 have a particularly simple expression for $\beta = 1$ (where we recover the exponential law), for $\beta = 2$, and for $\beta = \infty$ (constant lifetime). Table 6.1 indicates the values of m_1, m_2, and $\sigma_T^2 = m_2 - m_1^2$.

TABLE 6.1

	$\beta = 1$	$\beta = 2$	$\beta = \infty$
$\overline{T} = m_1$	$1/\lambda_0$	$\sqrt{\pi}/2\,\lambda_0$	$1/\lambda_0$
m_2	$2/\lambda_0^2$	$1/\lambda_0^2$	$1/\lambda_0^2$
$\sigma_T^2 = m_2 - m_1^2$	$1/\lambda_0^2$	$\left(1 - \frac{\pi}{4}\right)/\lambda_0^2$	0

We may transform the survival curve to a straight line by using the relation

$$\ln[-\ln v(t)] = \beta \ln t + \beta \ln \lambda_0 . \tag{6.22}$$

The Weibull distribution is the third asymptotic distribution of extreme values, in the terminology of Gumbel [26]. If we consider the random variable

$$Y = \min(X_1, X_2, \ldots, X_n), \tag{6.23}$$

where $X_1, \ldots, X_n$ are independent random variables all following the same Weibull law, the survival function of Y is

$$v_n(t) = \Pr\{ Y = \min(X_1, \ldots, X_n) > t \} = [v(t)]^n, \tag{6.24}$$

and relation (6.18) permits us to show easily that

$$v_n(t) = e^{-(n^{1/\beta}\lambda_0 t)^\beta}, \tag{6.25}$$

that is, that Y likewise follows a Weibull law in which the initial parameter λ_0 is replaced by $n^{1/\beta}\lambda_0$.

We note also that one may also use the Weibull law with a supplementary parameter $t_0 > 0$, introducing a shift in the zero of the survival curve

$$\begin{aligned} v(t) &= 1, & t &\leqslant t_0, \\ &= e^{-[\lambda_0(t-t_0)]^\beta}, & t &\geqslant t_0 . \end{aligned} \tag{6.25'}$$

(4) Log-Normal Law or the Law of Galton

(6.26) $$i(t) = \frac{1}{\sigma t\sqrt{2\pi}} e^{-(\ln t - \mu)^2/2\sigma^2}, \quad \mu \in \mathbf{R}, \quad \sigma \in \mathbf{R}_0^+, \quad t \geqslant 0,$$

(6.27) $$v(t) = \frac{1}{\sigma\sqrt{2\pi}} \int_t^\infty \frac{1}{u} e^{-(\ln u - \mu)^2/2\sigma^2} du,$$

(6.28) $$\lambda(t) = \frac{e^{-(\ln t - \mu)^2/2\sigma^2}}{t\int_t^\infty \frac{1}{u} e^{-(\ln u - \mu)^2/2\sigma^2} du}.$$

The log-normal law is the law of a random variable T whose natural logarithm[9]

(6.29) $$X = \ln T$$

follows a normal law (Laplace–Gauss law) with mean μ and variance σ^2; the variable

(6.30) $$Y = \frac{\ln T - \mu}{\sigma}$$

then follows a reduced centered normal law with probability density

(6.31) $$g(y) = \frac{1}{\sqrt{2\pi}} e^{-y^2/2}$$

and with complementary distribution function

(6.32) $$G(y) = \frac{1}{\sqrt{2\pi}} \int_y^\infty e^{-u^2/2} du.$$

The functions g and G allow a simpler representation of expressions (6.26)–(6.28):

(6.33) $$i(t) = \frac{1}{\sigma t} g\left(\frac{\ln t - \mu}{\sigma}\right),$$

(6.34) $$v(t) = G\left(\frac{\ln t - \mu}{\sigma}\right),$$

(6.35) $$\lambda(t) = \frac{g\left(\dfrac{\ln t - \mu}{\sigma}\right)}{\sigma t G\left(\dfrac{\ln t - \mu}{\sigma}\right)}.$$

[9] One sometimes defines the log-normal law using base-ten logarithms.

Figure 6.4a–c indicates the form of the probability density, the survival curve, and the failure rate, respectively.

The mean and the variance are

$$\overline{T} = e^{\mu + \sigma^2/2}, \tag{6.36}$$

$$\sigma_T^2 = e^{2\mu + \sigma^2}(e^{\sigma^2} - 1). \tag{6.37}$$

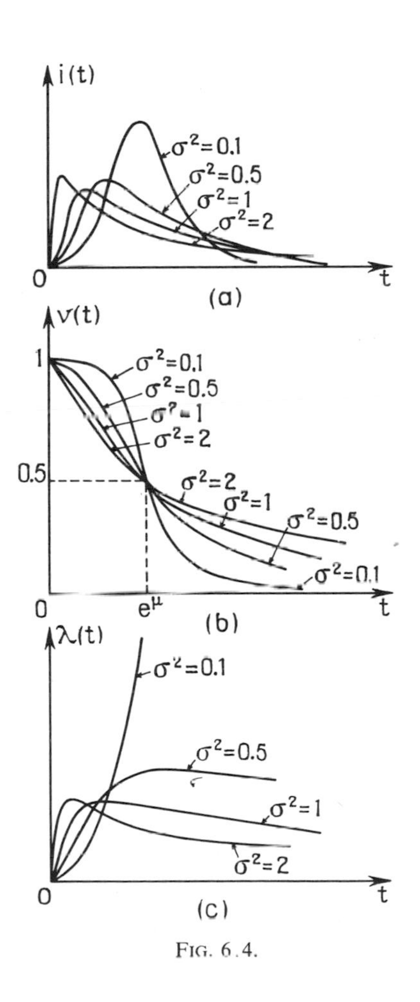

FIG. 6.4.

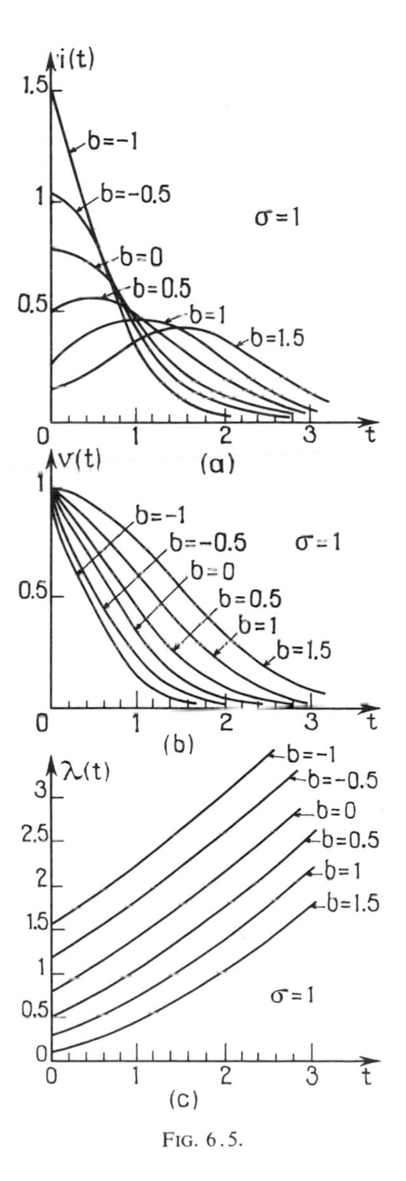

FIG. 6.5.

More generally, the noncentered moment of order r is given by

(6.38) $$m_r = e^{\mu r + r^2\sigma^2/2}.$$

It is interesting to note that the value of the median is e^μ; thus all the curves $v(t)$ corresponding to the same parameter μ pass through the point $(e^\mu; 0{,}5)$ whatever the value of σ (Fig. 6.4b). On the other hand, the value of the mode is $e^{\mu-\sigma^2}$; this decreases markedly as σ increases, and inversely, if $\sigma \to 0$, the mode tends toward e^μ (see Fig. 6.4a). The probability density attained at the mode is $1/(\sigma\sqrt{2\pi}\, e^{\mu-\sigma^2/2})$.

(5) TRUNCATED NORMAL LAW (Fig. 6.5)

(6.39) $$i(t) = \frac{1}{k\sigma\sqrt{2\pi}} e^{-(t-b)^2/2\sigma^2}, \quad b \in \mathbf{R},\ \sigma \in \mathbf{R}^+,\ t \geqslant 0;$$

where

(6.40) $$k = \frac{1}{\sigma\sqrt{2\pi}} \int_0^\infty e^{-(t-b)^2/2\sigma^2}\, dt,$$

(6.41) $$v(t) = \frac{1}{k\sigma\sqrt{2\pi}} \int_t^\infty e^{-(u-b)^2/2\sigma^2}\, du,$$

(6.42) $$\lambda(t) = \frac{e^{-(t-b)^2/2\sigma^2}}{\int_t^\infty e^{-(u-b)^2/2\sigma^2}\, du}.$$

If X follows a normal law with mean b and variance σ, the conditional law of X, knowing that $X \geqslant 0$, is the law above. The normal law is often used to represent the lifetime T of a component; but this law has the disadvantage, from the theoretical point of view, of giving a nonzero probability to the event $T < 0$, and use of the normal law truncated at the origin is more correct.

Using functions (6.31) and (6.32), we may rewrite the expressions above in the form

(6.43) $$i(t) = \frac{1}{k\sigma} g\left(\frac{t-b}{\sigma}\right),$$

(6.44) $$v(t) = \frac{1}{k} G\left(\frac{t-b}{\sigma}\right),$$

(6.45) $$\lambda(t) = \frac{g\left(\frac{t-b}{\sigma}\right)}{\sigma G\left(\frac{t-b}{\sigma}\right)},$$

where

$$k = G\left(-\frac{b}{\sigma}\right). \tag{6.46}$$

The mean and the variance are

$$\overline{T} = b + \frac{\sigma}{k} g\left(\frac{b}{\sigma}\right), \tag{6.47}$$

$$\sigma_T^2 = \frac{\sigma^2}{2k} - \frac{\sigma^2}{k\sqrt{\pi}} \Gamma_{b^2/2\sigma^2}\left(\frac{3}{2}\right) - \left[\frac{\sigma}{k} g\left(\frac{b}{\sigma}\right)\right]^2, \tag{6.48}$$

where

$$\Gamma_x(k) = \int_0^x u^{k-1} \, e^{-u} \, du$$

is the incomplete gamma function.

(6) GUMBEL'S LAW OR THE LAW OF EXTREME VALUES, TYPE I (Fig. 6.6)

$$i(t) = \beta\lambda_0 \, e^{\lambda_0 t - \beta(e^{\lambda_0 t} - 1)}, \quad \lambda_0, \beta \in \mathbf{R}^+, \tag{6.49}$$

$$v(t) = e^{-\beta(e^{\lambda_0 t} - 1)}, \tag{6.50}$$

$$\lambda(t) = \beta\lambda_0 \, e^{\lambda_0 t}. \tag{6.51}$$

The probability density $i(t)$ presents a mode for $t = (1/\lambda_0) \ln 1/\beta$, when $\beta < 1$; for $\beta > 1$, it decreases constantly. The failure rate is increasing. The median is equal to

$$\frac{1}{\lambda_0} \ln\left(1 + \frac{1}{\beta} \ln 2\right). \tag{6.52}$$

The moments of this law are not expressible in a simple fashion.

If the random variable X follows an exponential law with parameter β, the variable

$$T = \frac{1}{\lambda_0} \ln (X + 1)$$

follows the law of Gumbel (note that $X \geqslant 0$ implies $T \geqslant 0$). One may, moreover, generalize the law (6.50) by starting with a variable X that follows a Weibull law (of which the exponential law is a particular case).

The law of Gumbel is the first asymptotic distribution of extremal values (modified for convenience to a nonnegative random variable) in the classification of Gumbel [26]. One may easily verify that it possesses the same property as does Weibull's law (see (6.23)–(6.25)): the minimum of n independent

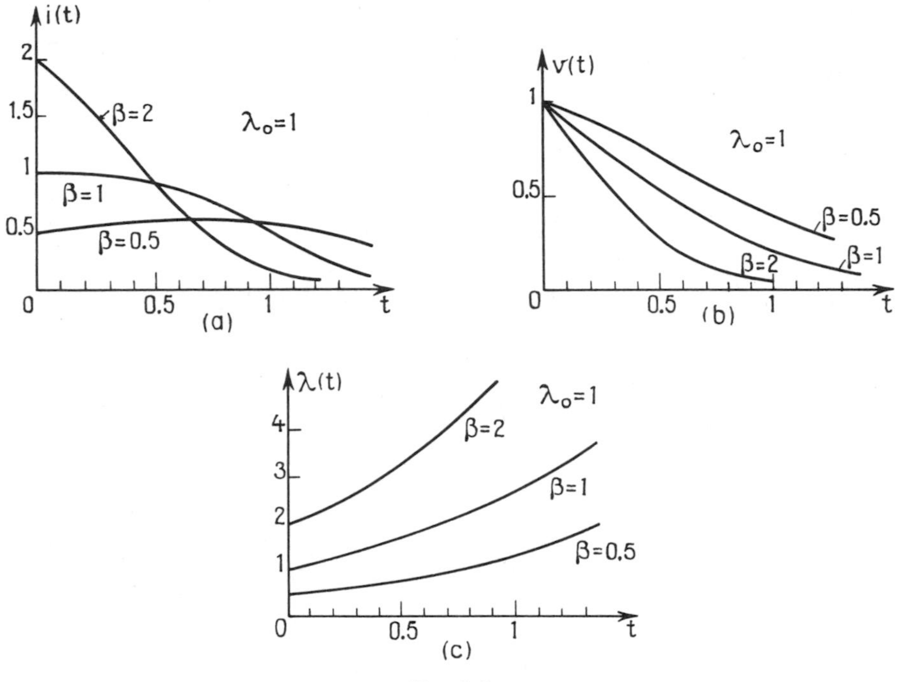

FIG. 6.6.

variables all following the same law of Gumbel with parameter β follows a law of Gumbel with parameter $n\beta$.

Laws of Type I. We now suppose that the lifetime of a component is a discrete quantity; more precisely we suppose that the n values that may be taken by T belong to $\mathbf{N} = \{0, 1, 2, 3, \dots\}$.

The principal laws of type I used as survival laws are:

binomial law,
Poisson law,
geometric law,
negative binomial law.

The probability that $T = n$ will be designated by $p(n)$, the complementary distribution function by $v(n)$, and the failure rate by $p_c(n)$:

$$p(n) = \text{pr}\{T = n\}, \tag{6.53}$$

$$v(n) = \text{pr}\{T > n\}, \tag{6.54}$$

$$p_c(n) = \frac{p(n+1)}{v(n)}. \tag{6.55}$$

We shall rapidly review the formulas concerning the first three of these laws, since these are very well known, and dwell a little more on the fourth since it is less so.

BINOMIAL LAW

(6.56) $$p(n) = \binom{m}{n} p^n q^{m-n}, \quad n \in \{0, 1, 2, \ldots, m\},$$
$$0 < p < 1,$$
$$= 0, \quad n > m, \quad q = 1 - p,$$
$$m \in \mathbf{N}_0 .$$

(6.57) $$v(n) = \sum_{i=n+1}^{m} \binom{m}{i} p^i q^{m-i}, \quad n \in \{0, 1, 2, \ldots, m-1\},$$
$$= 0, \quad n \geqslant m .$$

(6.58) $$p_c(n) = \frac{\binom{m}{n+1} p^{n+1} q^{m-n-1}}{\sum_{i=n+1}^{m} \binom{m}{i} p^i q^{m-i}}, \quad n \in \{-1, 0, 1, 2, \ldots, m-1\} .$$

(6.59) $$\overline{T} = E[T] = mp ,$$

(6.60) $$\sigma_T^2 = mpq .$$

LAW OF POISSON

(6.61) $$p(n) = \frac{\lambda_0^n \, \mathrm{e}^{-\lambda_0}}{n\,!}, \quad n \in \mathbf{N}, \lambda_0 \in \mathbf{R}^+ ,$$

(6.62) $$v(n) = \mathrm{e}^{-\lambda_0} \sum_{i=n+1}^{\infty} \frac{\lambda_0^i}{i\,!},$$

(6.63) $$p_c(n) = \frac{\lambda_0^{n+1}/(n+1)\,!}{\sum_{i=n+1}^{\infty} \lambda_0^i / i\,!},$$

(6.64) $$\overline{T} = E[T] = \lambda_0 ,$$

(6.65) $$\sigma_T^2 = \lambda_0 .$$

GEOMETRIC LAW (OR PASCAL'S LAW)

(6.66) $$p(n) = pq^n, \quad n \in \mathbf{N}, 0 < p < 1, q = 1 - p ,$$

(6.67) $$v(n) = q^{n+1} ,$$

(6.68) $$p_c(n) = p .$$

Thus the failure rate for this law is constant. This law plays the same role for survival laws defined in $\mathbf{N}$ as does the exponential law defined in $\mathbf{R}^+$.

We also have

$$\overline{T} = E[T] = \frac{q}{p}, \tag{6.69}$$

$$\sigma_T^2 = \frac{q}{p^2}. \tag{6.70}$$

The geometric law is generalized by the "hypergeometric law"

$$p(n) = \frac{\binom{mp}{n} \cdot \binom{mq}{r-n}}{\binom{m}{r}}, \qquad \begin{aligned} &n \in \{0, 1, 2, \ldots, r\}, \\ &r \in \{0, 1, 2, \ldots, m\}, \\ &p \in \left\{0, \frac{1}{m}, \frac{2}{m}, \ldots, 1\right\}, \\ &q = 1 - p. \end{aligned} \tag{6.71}$$

Negative Binomial Law We dwell a little more on this law that is less well known and which plays an important role in various applications. This law is defined by

$$p(n) = \binom{r + n - 1}{n} p^r q^n, \qquad \begin{aligned} &n \in \mathbf{N}, \\ &r \in \mathbf{R}_0^+, \\ &0 < p < 1, \\ &q = 1 - p, \end{aligned} \tag{6.72}$$

where $\binom{r+n-1}{n}$ is defined, even for r not an integer, by the usual relation

$$\binom{x}{n} = \frac{x(x-1)\ldots(x-n+1)}{n!}, \qquad \begin{aligned} &x \in \mathbf{R}, \\ &n \in \mathbf{N}. \end{aligned} \tag{6.73}$$

By using the notation[10]

$$\binom{-r}{n} = (-1)^n \binom{r + n - 1}{n} \tag{6.74}$$

[10] This notation follows from the following property. One has, by definition (6.73),

$$\binom{\nu}{k} = \frac{\nu(\nu - 1)\ldots(\nu - k + 1)}{k!}.$$

Nothing prevents us from using this formula for negative ν (with the reservation of no longer giving to $\binom{\nu}{k}$ its combinatoric calculus meaning, where this quantity represents the number of combinations of ν objects taken k at a time). By exchanging ν for $-\nu$, one then has for $\nu > 0$

$$\binom{\nu}{k} = \frac{(-\nu)(-\nu - 1)\ldots(-\nu - k + 1)}{k!} = (-1)^k \frac{\nu(\nu + 1)\ldots(\nu + k - 1)}{k!} = (-1)^k \binom{\nu + k - 1}{k}.$$

we may also write

$$p(n) = \binom{-r}{n} p^r(-q)^n . \tag{6.75}$$

We shall then have

$$v(n) = p^r \sum_{i=n+1}^{\infty} \binom{-r}{i} (-q)^i . \tag{6.76}$$

We may also put $v(n)$ in the following form, which has the appearance of a binomial law:

$$v(n) = \sum_{i=0}^{r-1} \binom{n+r}{i} p^i q^{r+n-i} . \tag{6.77}$$

We also have

$$p_c(n) = \frac{\binom{-r}{n+1} p^r(-q)^{n+1}}{p^r \sum_{i=n+1}^{\infty} \binom{-r}{i}(-q)^i} , \tag{6.78}$$

$$\bar{T} = E[T] = \frac{rq}{p} , \tag{6.79}$$

$$\sigma_T^2 = r \frac{q}{p^2} . \tag{6.80}$$

Recall that, for r an integer, the negative binomial law is the law of the number n of failures encountered before the rth success in a sequence of Bernoulli trials (that is, repeated, independent trials where the two possible results are success or failure), p being the probability of success in a trial. One may show [23] that the sum of r independent variables, each distributed according to the same geometric law, follows a negative binomial law. In other words, if

$$N = N_1 + N_2 + \cdots + N_r \tag{6.81}$$

where the variables N_i, $i = 1, 2, \ldots, r$, all follow the law

$$p(n_i) = pq^{n_i} , \tag{6.82}$$

then N follows the law

$$p(n) = \binom{r+n-1}{n} p^r q^n . \tag{6.83}$$

This property allows one to obtain moments (6.79) and (6.80) of the negative binomial law with respect to the corresponding moments (6.69) and (6.70) of the geometric law. Likewise, (6.77) may be obtained by noting that $v(n)$ is the probability that the first $n + r$ trials give at least r successes.

7 Survival Law of Nonnew Equipment[11]

Suppose that the equipment put into service has age a. The conditional survival law is then no longer $v(t)$, but another law that will be designated $v_a(t)$. We now proceed to show how to determine this function.

The a priori probability that a new component will attain the age $a + \tau$ without deterioration is $v(a + \tau)$. This probability may be written as

$$v(a + \tau) = v(a) . v_a(\tau) , \tag{7.1}$$

or

$$v_a(\tau) = \frac{v(a + \tau)}{v(a)} . \tag{7.2}$$

Thus the survival curve of a component having initial wear (already used until age a) is obtained by shifting the survival curve to the left through a and multiplying the ordinates of the curve obtained by $1/v(a)$ (Fig. 7.1). One ought not to be surprised that for certain values of t, the probability $v_a(t)$ may be greater than the probability $v(t)$; it all depends on the nature of the survival law.

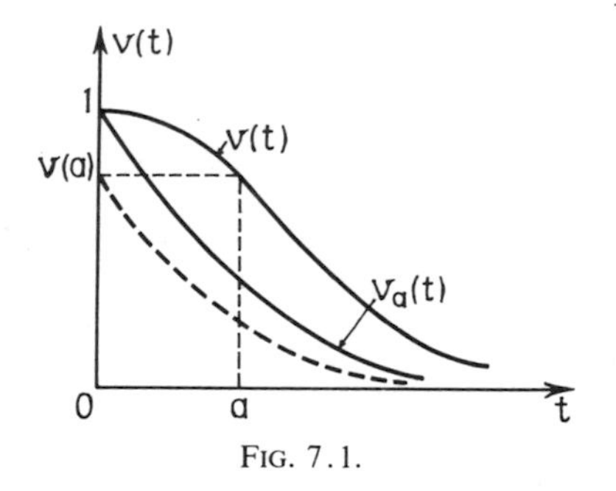

FIG. 7.1.

In the case of a law of type II, relation (4.13) immediately gives an expression for $v_a(\tau)$ as a function of the failure rate:

$$v_a(\tau) = \exp\left(- \int_a^{a+\tau} \lambda(u)\, du\right). \tag{7.3}$$

In other words, the cumulative failure rate $\Lambda_a(\tau)$ of nonnew equipment is

$$\Lambda_a(\tau) = \Lambda(a + \tau) - \Lambda(a) , \tag{7.4}$$

and its instantaneous failure rate is

$$\lambda_a(\tau) = \lambda(a + \tau) . \tag{7.5}$$

[11] Use of the word *used* would be very ambiguous.

The failure rate curve therefore remains the same to within a translation. In particular, if the survival curve of the new equipment has an increasing failure rate (cf. Section 10), then so does the nonnew equipment.

The exponential law possesses an interesting property:

$$(7.6) \qquad v_a(\tau) = \frac{e^{-\lambda_0(a+\tau)}}{e^{-\lambda_0 a}} = e^{-\lambda_0 \tau}.$$

A component whose survival law is the exponential law with a failure rate λ_0 obeys this law whatever its age when put into service. One may easily show that the exponential law is the only law that has this property.

8 Survival Law with Guarantee. Survival Law with a Limit on Functioning

Certain kinds of equipment have a guarantee. We suppose that, in the interval $[0, a[$, any equipment that fails will be repaired without cost and may be put back into service with the same degree of use; the equipment put into place thus has the same age as that which failed. In this case curves $v(t)$ and $i(t)$ are modified in the following fashion.

From $t = 0$ to $t = a$, failures may be neglected since the equipment is repaired without cost[12]; thus denoting the survival curve with guarantee by $v_g(t; a)$ we have

$$(8.1) \qquad v_g(t; a) = 1, \qquad 0 \leqslant t < a.$$

At the date $t = a - \varepsilon$ all the pieces of equipment have age $a - \varepsilon$, and their survival curve is then the curve $v_{a-\varepsilon}(t)$ shifted to the right through the value[13] $t = a - \varepsilon$. Finally (Fig. 8.1),

$$(8.2) \qquad v_g(t; a) = \frac{v(t)}{v(a^-)}, \qquad t \geqslant a.$$

where

$$v(a^-) = \lim_{\varepsilon \to 0} v(a - \varepsilon).$$

[12] Note, however, that failures occurring during the guarantee period may entail some unavailability of the material, which may be extremely inconvenient.

[13] The result of Section 7 supposes that the equipment is in a good state at age a. The guarantee excluding a failure that occurs exactly at age a, we need only use this result for $a - \varepsilon$, where $\varepsilon \to 0$.

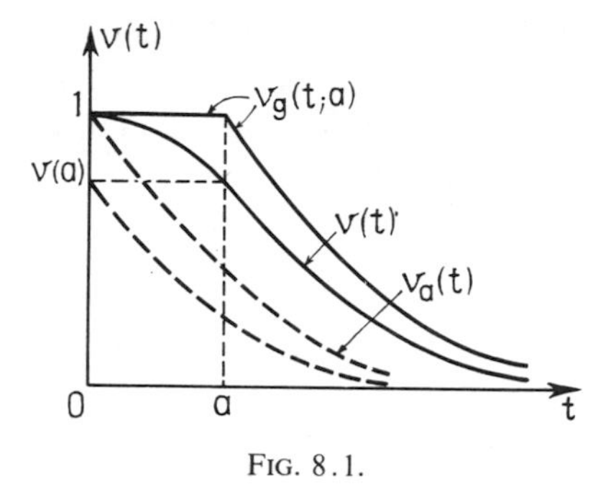

FIG. 8.1.

Thus

(8.3)
$$v_g(t\,;a) = 1\,, \qquad 0 \leqslant t < a\,,$$
$$= \frac{v(t)}{v(a^-)}\,, \qquad t \geqslant a\,.$$

One may imagine a more general case where the guarantee extends to the interval $[a, b[$. The formulas below give the corresponding functions, which are associated with Figs. 8.2a–c. We are supposing as above that the repaired equipment conserves its age[14]:

(8.4)
$$v_g(t) = v(t)\,, \qquad 0 \leqslant t < a\,,$$
$$= v(a^-)\,, \qquad a \leqslant t < b\,,$$
$$= \frac{v(a^-)}{v(b^-)}\,v(t)\,, \qquad t \geqslant b\,.$$

In the case of a survival law of type IIa, we also have

(8.5)
$$i_g(t) = i(t)\,, \qquad 0 \leqslant t < a,$$
$$= 0\,, \qquad a \leqslant t < b\,,$$
$$= \frac{v(a^-)}{v(b^-)}\,i(t)\,, \qquad t \geqslant b\,,$$

(8.6)
$$\lambda_g(t) = \lambda(t)\,, \qquad 0 \leqslant t < a, \text{ and } \quad t \geqslant b\,,$$
$$= 0, \qquad a \leqslant t < b\,.$$

Another interesting case is that in which we consider a *survival law with a limit on functioning* (Fig. 8.3). We are given in this case a limit on functioning θ; and the equipment is put out of service at age θ, if it attains this age; the survival law is therefore modified. Call this new law $v_h(t;\,\theta)$. We have

(8.7)
$$v_h(t\,;\theta) = v(t)\,, \qquad 0 \leqslant t < \theta\,,$$
$$= 0\,, \qquad t \geqslant \theta,$$

[14] In the case of a law of type I, we have
$$v(a^-) = v(a-1)\,.$$

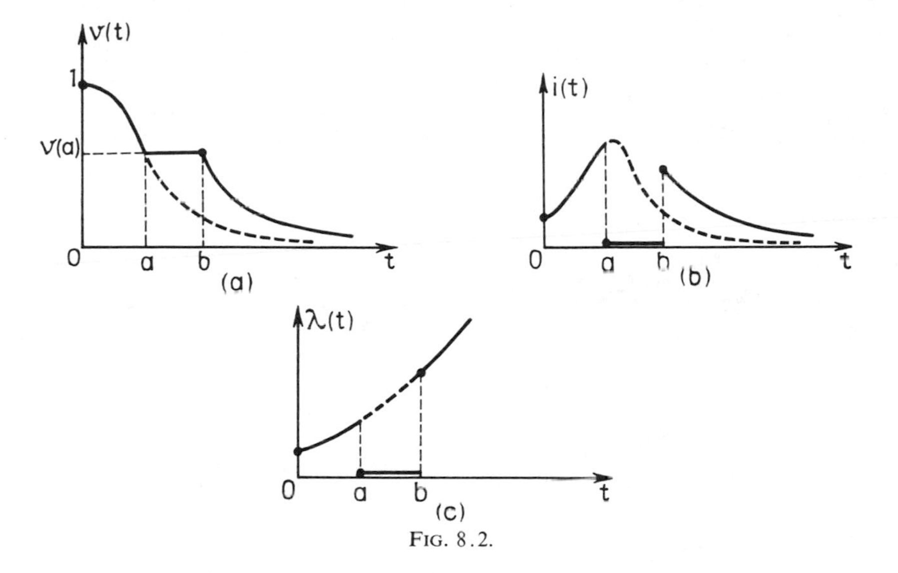

FIG. 8.2.

and

(8.8)
$$\begin{aligned} i_h(t\,;\theta) &= i(t)\,, \qquad 0 \leqslant t < \theta\,, \\ &= \delta(t-\theta).v(\theta)\,, \qquad t = \theta\,, \\ &= 0\,, \qquad t > \theta\,, \end{aligned}$$

where $\delta(t)$ is the Dirac measure, or Dirac's delta function. The probability density at $t = \theta$ is thus infinite; it is the same evidently for the failure rate. According to (8.7) and (4.19), the cumulative failure rate is given by

(8.9)
$$\begin{aligned} \Lambda_h(t) &= \Lambda(t)\,, \qquad 0 \leqslant t < \theta, \\ &= \infty\,, \qquad t \geqslant \theta\,. \end{aligned}$$

The limit on functioning θ is often called the "removal age."

The mean lifetime may easily be obtained in the same fashion as in (5.10):

(8.10)
$$\begin{aligned} \overline{T}_h(\theta) &= -\int_0^\infty t\,\mathrm{d}v_h(t\,;\theta) \\ &= -\int_0^\theta t\,\mathrm{d}v(t) + \theta v(\theta) \\ &= -\,[tv(t)]_0^\theta + \int_0^\theta v(t)\,\mathrm{d}t + \theta v(\theta) \\ &= \int_0^\theta v(t)\,\mathrm{d}t = \int_0^\infty v(t)\,\mathrm{d}t - \int_\theta^\infty v(t)\,\mathrm{d}t \\ &= \overline{T} - w(\theta)\,. \end{aligned}$$

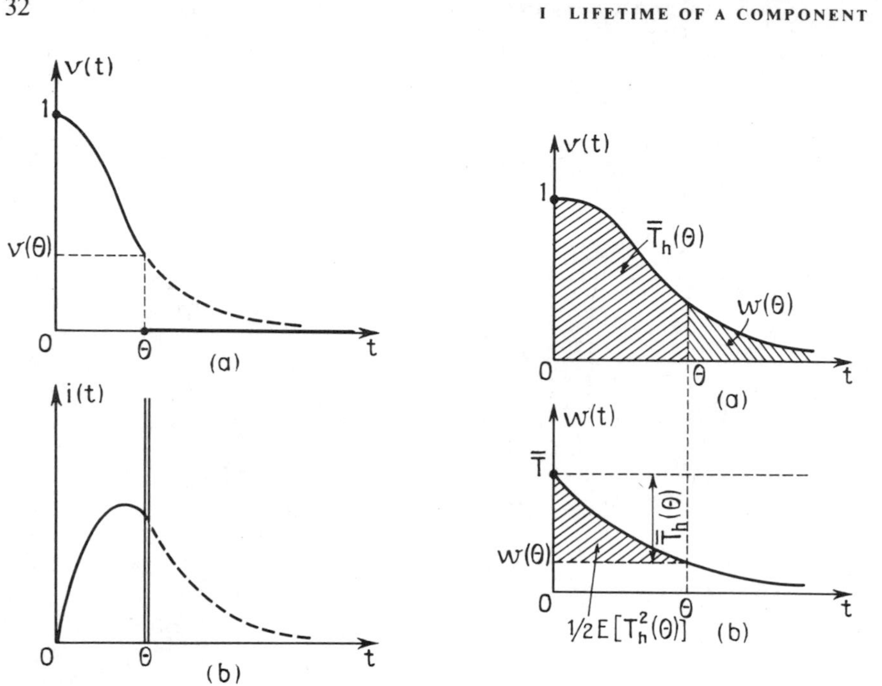

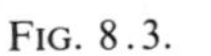
FIG. 8.3.

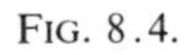
FIG. 8.4.

By fixing a limit on functioning θ we thus diminish the mean life of a quantity $w(\theta)$, where the function $w(t)$ is defined by (5.7).

The moment of second order is

$$(8.11) \qquad E[T_h^2(\theta)] = -\int_0^\theta t^2 \, \mathrm{d}v(t) + \theta^2 \, v(\theta)$$

$$= -\,[t^2 \, v(t)]_0^\theta + 2\int_0^\theta tv(t) \, \mathrm{d}t + \theta^2 \, v(\theta)$$

$$= -\,2\int_0^\theta t \, \mathrm{d}w(t)$$

$$= -\,2\,\theta w(\theta) + 2\int_0^\theta w(t) \, \mathrm{d}t \, .$$

Figure 8.4 gives a geometric interpretation of the various quantities above.

The variance may be obtained classically from

$$(8.12) \qquad \sigma^2_{T_h(\theta)} = E[T_h^2(\theta)] - (\overline{T}_h(\theta))^2 \, .$$

CHAPTER II

EQUIPMENT WITH AN INCREASING FAILURE RATE

9 Introduction

In this chapter we shall study a particular class of survival functions, characterized by the property that the failure rate increases with the age of the equipment, or at least is nondecreasing. One may in fact expect that aging of equipment increases the probability that it will fail. It is, however, necessary to make two remarks:

(1) One often observes, at the beginning of the life of a piece of equipment, some failures "of youth"; equipment that has successfully passed this point then presents a reduced failure rate. This is why one often gives as a more general failure curve the "basin" of Fig. 9.1.

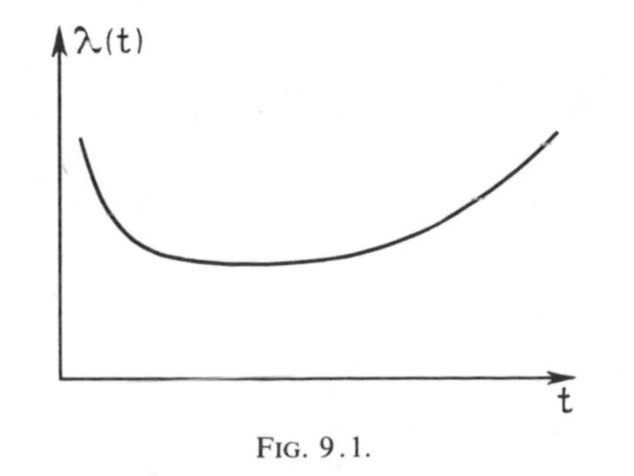

FIG. 9.1.

(2) The effect of aging may be relevant only at a very late age, having a very low probability of being attained. It will thus not be observed in practice, the failures almost all being produced in the "flat" part of the theoretical "basin" curve. This seems to be the case with the great majority of electronic components, for which one usually supposes an exponential survival law.

It is, however, useful to examine the particular properties of survival curves for nondecreasing failure rates, for which the exponential curve (constant failure rate) constitutes a limiting case. We shall see that certain of these properties persist for a slightly larger class of survival functions, having a failure rate that is nondecreasing "in the mean."

10 Survival Functions with Increasing (Decreasing) Failure Rate

We first review the very important notion of failure rate, which has been defined in Section 4:

(10.1)
$$\lambda(t) = \frac{i(t)}{v(t)} = -\frac{v'(t)}{v(t)} .$$

We now define the *failure rate in an interval* $]t, t + x]$, $x > 0$, by the expression

(10.2)
$$\mu(t\,; x) = \frac{v(t) - v(t + x)}{v(t)} = 1 - v_t(x)$$

where $v_t(x)$ is the survival law of a piece of equipment with initial age t (see (7.2)). In the case of a law of type I we shall have the same definition:

(10.3)
$$\mu(n\,; u) = \frac{v(n) - v(n + u)}{v(n)}, \qquad n \in \mathbf{N}, \quad u \in \mathbf{N}_0 .$$

The failure rate in an interval is related to the cumulative failure rate and to the instantaneous failure rate by the following relation, obtained by expressing $v(t)$ as a function of $\Lambda(t)$ through (4.19),

(10.4)
$$\mu(t\,; x) = 1 - \exp(-[\Lambda(t + x) - \Lambda(t)]) = 1 - \exp\left(-\int_t^{t+x} \lambda(u)\, du\right).$$

One the other hand, for a law of type IIa we have

(10.5)
$$\lambda(t) = \lim_{x \to 0} \frac{\mu(t\,; x)}{x}$$

and, for a law of type I,

(10.6)
$$p_c(n) = \mu(n\,; 1) .$$

One should note that, in the case of a law of type IIa, the instantaneous failure rate has the dimensions of the reciprocal of age, and that the failure rate in an interval is without dimension.

Definition I (concerning survival laws of type IIa). *Survival function with increasing failure rate (IFR) (respectively, decreasing failure rate (DFR)). A survival function $v(t)$ will be said to be IFR (respectively, DFR) if and only if*

$$(10.7) \qquad \forall t_1, t_2 \in \mathbf{R}^+ : (t_2 > t_1) \Rightarrow (\lambda(t_2) \geqslant \lambda(t_1)) \qquad (\text{resp.} \leqslant),$$

that is, if $\lambda(t)$ is a nondecreasing[1] *(respectively, nonincreasing) function.*

This definition is equivalent to the following if $\lambda(t)$ is differentiable:

$$(10.8) \qquad \forall t \in \mathbf{R}^+ : \lambda'(t) \geqslant 0 \qquad (\text{resp.} \leqslant)$$

where

$$\lambda'(t) = \frac{\mathrm{d}}{\mathrm{d}t}\lambda(t).$$

Definition II (concerning survival laws of type I). *Survival function with increasing failure rate (IFR) (respectively, decreasing failure rate (DFR)). A survival function $v(n)$ will be said to be IFR (respectively, DFR) if and only if*

$$(10.9) \qquad \forall n_1, n_2 \in \mathbf{N} : (n_2 > n_1) \Rightarrow (p_c(n_2) \geqslant p_c(n_1)) \qquad (\text{resp.} \leqslant),$$

that is, if $p_c(n)$ is a nondecreasing (respectively, nonincreasing) function for $n \geqslant 0$.

Another definition of IFR or DFR functions. *A survival function $v(t)$ is IFR if and only if*

$$(10.10) \qquad \forall t \in \mathbf{R} \text{ and such that } v(t) > 0, \text{ and } \forall x \in \mathbf{R}^+ :$$

$$\mu(t;x) = \frac{v(t) - v(t+x)}{v(t)}$$

is a nondecreasing function of t (respectively, DFR if $\mu(t;x)$ is nonincreasing in $\mathbf{R}^+$), for t an integer in the case of a law of type I.

This definition with respect to failure rate by intervals has the advantage

[1] In order to avoid any ambiguity due to the terminology employed, we shall use the following definitions: Let $f(x)$ be a function defined in $[a, b]$, $b > a$; then if $\forall x_1, x_2 \in [a, b]$:

$(x_2 > x_1) \Rightarrow (f(x_2) \geqslant f(x_1))$: the function is nondecreasing,
$(x_2 > x_1) \Rightarrow (f(x_2) > f(x_1))$: the function is increasing (we also say, strictly increasing),
$(x_2 > x_1) \Rightarrow (f(x_2) \leqslant f(x_1))$: the function is nonincreasing,
$(x_2 > x_1) \Rightarrow (f(x_2) < f(x_1))$: the function is decreasing (or strictly decreasing),
$(x_2 > x_1) \Rightarrow (f(x_2) = f(x_1))$: the function is constant; it is also nondecreasing and nonincreasing according to the above definitions.

of being applicable to survival functions of type I as well as type II (including type IIb).

We shall prove the equivalence of this definition with (10.7) for an IFR function of type I; one proceeds similarly for a DFR function.

First we remark that if (10.10) is nondecreasing, then so is the conditional failure rate, according to (10.6). Conversely, suppose that $p_c(n)$ is nondecreasing. According to definition (4.11), we have

$$p_c(n) = \frac{v(n) - v(n+1)}{v(n)} = 1 - \frac{v(n+1)}{v(n)} \tag{10.11}$$

or

$$\frac{v(n+1)}{v(n)} = 1 - p_c(n)\,,$$

from which we easily obtain

$$v(n) = v(0) \prod_{i=0}^{n-1} [1 - p_c(i)] = \prod_{i=-1}^{n-1} [1 - p_c(i)]\,. \tag{10.12}$$

On the other hand,

$$\mu(n\,;h) = \frac{v(n) - v(n+h)}{v(n)} = 1 - \frac{v(n+h)}{v(n)}\,, \tag{10.13}$$

which may be expressed, using (10.12), as

$$\mu(n\,;h) = 1 - \prod_{i=n}^{n+h-1} [1 - p_c(i)]\,. \tag{10.14}$$

The condition that $p_c(i)$ is nondecreasing then implies that $1 - p_c(i)$ is nonincreasing, and therefore that the product appearing in (10.14) is similarly nonincreasing; $\mu(n;\,h)$ is thus nondecreasing.

We now pass to the case of a law of type IIa. Relation (10.5) shows immediately that (10.10) implies that $\lambda(t)$ is nondecreasing. We therefore prove the reverse implication.

Suppose that

$$(t_2 > t_1) \Rightarrow (\lambda(t_2) \geqslant \lambda(t_1))\,; \tag{10.15}$$

then for $t_2 > t_1$ and for all $\tau \in \mathbf{R}^+$,

$$\lambda(\tau + t_2) \geqslant \lambda(\tau + t_1)\,, \tag{10.16}$$

from which

$$\int_0^\tau \lambda(\theta + t_2)\, d\theta \geqslant \int_0^\tau \lambda(\theta + t_1)\, d\theta\,. \tag{10.17}$$

Then, by changing variables,

$$\int_{t_2}^{t_2+\tau} \lambda(\alpha)\, d\alpha \geqslant \int_{t_1}^{t_1+\tau} \lambda(\alpha)\, d\alpha\,; \tag{10.18}$$

from which

$$1 - \exp\left(-\int_{t_2}^{t_2+\tau} \lambda(\alpha)\, d\alpha\right) \geqslant 1 - \exp\left(-\int_{t_1}^{t_1+\tau} \lambda(\alpha)\, d\alpha\right). \tag{10.19}$$

Thus we have, according to (10.4),

$$\mu(t_2\,;\tau) \geqslant \mu(t_1\,;\tau)\,. \tag{10.20}$$

This shows us that $(v(t) - v(t + x))/v(t)$ is nondecreasing in t.

Case of Survival Laws of Type IIb. We have remarked above that definition (10.10) is applicable to laws of type IIb, for which the failure rate $\lambda(t)$ is not everywhere defined. We shall go on to see, however, that only a particular kind of survival law of type IIb may be IFR. In fact, condition (10.10) signifies that the function $v(t + x)/v(t)$ is nonincreasing; it is then the same for the function $v(u)/v(u - x)$, obtained by setting $u = t + x$. Suppose that $v(u)$ has a discontinuity at the point $u = \theta$, that is, that

$$\frac{v(\theta)}{v(\theta - x)} \leqslant \alpha < 1 \tag{10.21}$$

(recall that $v(t)$ is continuous on the right and nonincreasing). Then let $u > \theta$, and put

$$x = \frac{u - \theta}{n}\,, \tag{10.22}$$

where n is arbitrarily large. We may write

$$v(u) = \frac{v(u)}{v(u - x)} \cdot \frac{v(u - x)}{v(u - 2\,x)} \cdots \frac{v(\theta)}{v(\theta - x)} \cdot v(\theta - x) \tag{10.23}$$

by noting that $\theta = u - nx$. The condition that $v(u)/v(u - x)$ be nonincreasing shows that, applying (10.21), all the terms appearing in (10.23) are at least equal to α, and thus that

$$v(u) \leqslant \alpha^{n+1}\, v(\theta - x)\,. \tag{10.24}$$

Since n is as large as one wishes and α less than 1, this is possible only if

$$v(u) = 0\,, \qquad \forall u > \theta\,; \tag{10.25}$$

$v(t)$ being continuous on the right, one also has $v(\theta) = 0$. We therefore see

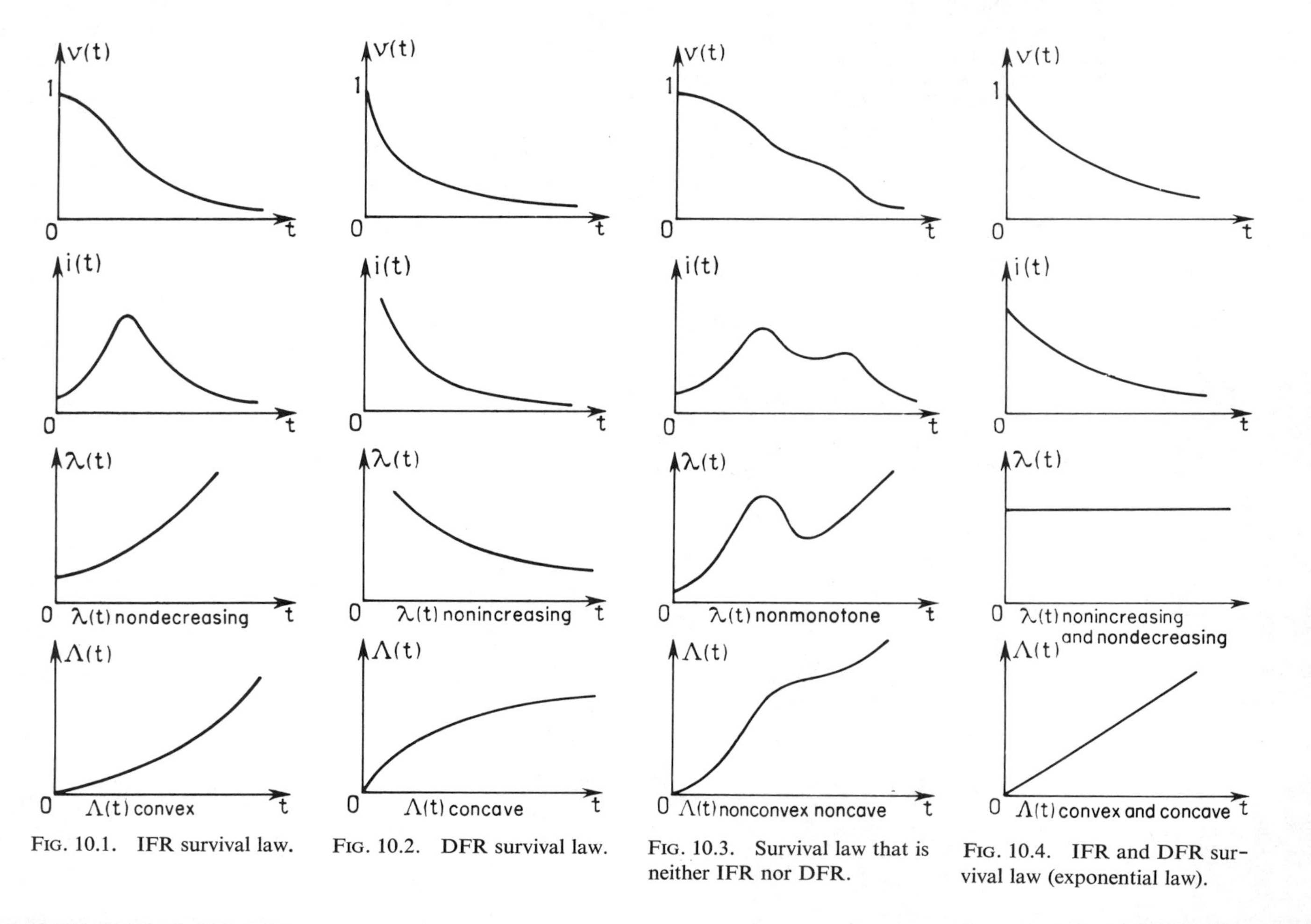

Fig. 10.1. IFR survival law.

Fig. 10.2. DFR survival law.

Fig. 10.3. Survival law that is neither IFR nor DFR.

Fig. 10.4. IFR and DFR survival law (exponential law).

that, if θ is a point of discontinuity of $v(u)$, we have $v(\theta) = 0$. The only laws of type IIb that may be IFR are thus those that may be obtained from a law of type IIa by introducing a limit on functioning (see Section 8).

Theorem 10.1. *A survival law $v(t)$ of type II is IFR if and only if the cumulative failure rate defined by (4.20) is convex*[2] *in the interval where it is defined, that is, for $v(t) > 0$,*

$$\Lambda(t) = -\ln v(t) \qquad \textit{is convex.} \tag{10.26}$$

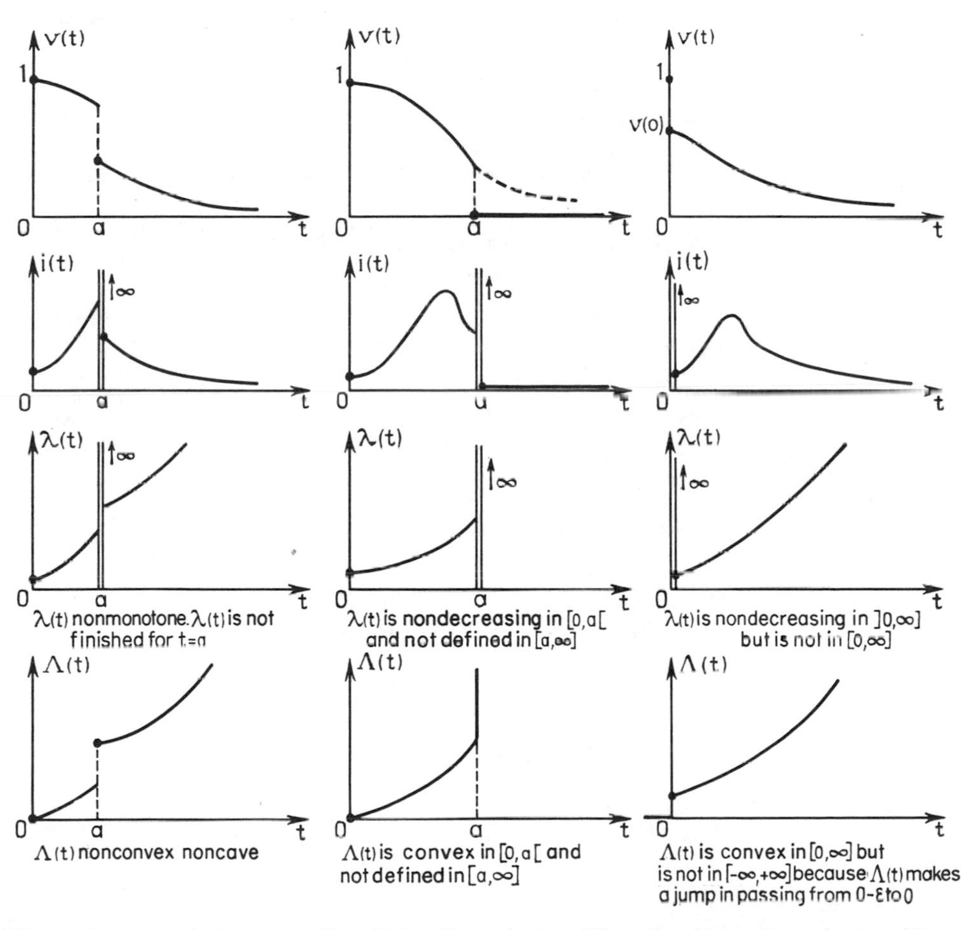

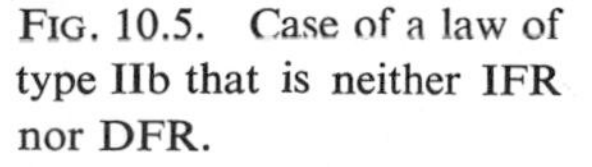

FIG. 10.5. Case of a law of type IIb that is neither IFR nor DFR.

FIG. 10.6. Case of a type IIb law that is IFR.

FIG. 10.7. Case of a type IIb law that is not IFR.

[2] Recall that a function $f(x)$ is convex if, for $0 \leqslant \alpha \leqslant 1$, we have $f[\alpha x_1 + (1-\alpha)x_2] \leqslant \alpha f(x_1) + (1-\alpha)f(x_2)$. A convex function is necessary continuous and has at every point a left derivative and a right derivative, which are nondecreasing. The second derivative, if it exists, is nonnegative. A concave function $f(x)$ is a function such that $-f(x)$ is convex.

In fact, we have seen that an IFR function has an instantaneous failure rate $\lambda(t)$ that is nondecreasing. The cumulative failure rate

$$\Lambda(t) = \int_0^t \lambda(u)\, du$$

is thus convex. Conversely, if $\Lambda(t)$ is convex, then $\lambda(t)$ is nondecreasing.

Functions whose logarithm is concave (or whose inverse is convex, which amounts to the same thing) are used in various branches of mathematics (see Barlow and Proschan [5]) under the name Pólya functions of order 2. In Appendix A one may find several facts about these functions, and also about totally positive functions of order 2 which they generalize.

IFR survival functions are those whose cumulative failure rate is concave in $[0, +\infty]$. In Figs. 10.1–10.7 various examples of survival functions that are IFR, DFR, or neither IFR nor DFR are shown.

11 Properties of IFR Functions

IFR functions have some important properties which have been studied in detail by Barlow and Proschan (see, in particular, Ref. [5]). DFR functions have analogous properties, but these functions are less useful in practice, and we shall content ourselves with mentioning their properties in passing. The proofs that we give are inspired by those of Barlow and Proschan, but are more simple in the majority of cases.

The exponential survival law

$$v(t) = e^{-\lambda_0 t}, \tag{11.1}$$

which has a constant failure rate, is at the same time IFR and DFR. It marks the boundary between these two families of survival laws. The theorems below exploit this property through bounding an IFR survival law and its moments by analogous quantities relative to the exponential law.

Theorem 11.1. *If the survival function $v(t)$ is IFR:*

(a) *either there exists a λ_0 such that $v(t) = e^{-\lambda_0 t}$ for all t, or*
(b) *for all $\lambda_0 > \lambda(0)$ there exists a $t_0 > 0$ such that*

$$v(t) > v_e(t), \qquad 0 < t < t_0, \tag{11.2}$$

$$v(t_0) = v_e(t_0), \tag{11.3}$$

$$v(t) < v_e(t), \qquad t > t_0, \tag{11.4}$$

where

$$v_e(t) = e^{-\lambda_0 t}, \tag{11.5}$$

and where $\lambda(0)$ is the failure rate at the origin of the law $v(t)$.

The theorem becomes evident if one considers the cumulative failure rates rather than the survival laws; recall (4.18) and (4.19):

(11.6) $$\Lambda(t) = \int_0^t \lambda(u)\, du$$

and

(11.7) $$v(t) = e^{-\Lambda(t)} .$$

The equation

(11.8) $$v(t) = v_e(t)$$

may thus be written

(11.9) $$\Lambda(t) = \Lambda_e(t) ,$$

where

(11.10) $$\Lambda_e(t) = \lambda_0 \, t$$

is the cumulative failure rate of the exponential law and $\Lambda(t)$ is by hypothesis a convex function whose derivative at the origin is $\lambda(0)$. If $\Lambda(t)$ is not a straight line, that is, if $\Lambda(t)$ is strictly convex at least in certain intervals, any line $\Lambda_e(t)$ of slope $\lambda_0 > \lambda(0)$ intersects once and only once (apart from the origin) the curve $\Lambda(t)$ (Fig. 11.1).

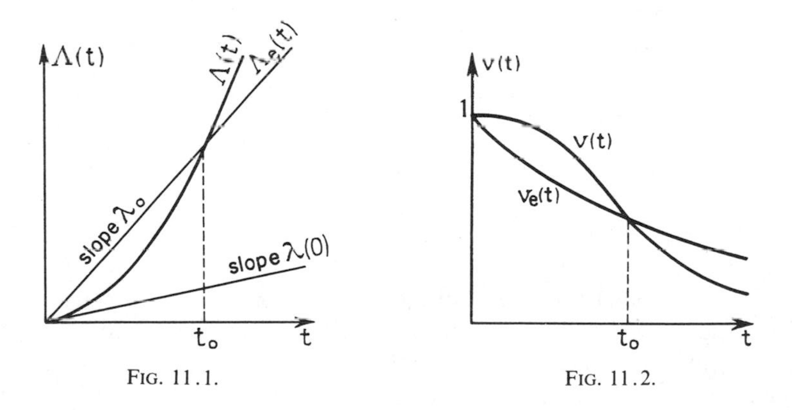

FIG. 11.1.

FIG. 11.2.

The respective positions of the curves $\Lambda(t)$ and $\Lambda_e(t)$ then imply, by virtue of (11.7), an inverse arrangement of curves $v(t)$ and $v_e(t)$ (Fig. 11.2), which proves the theorem.

In the case of a DFR function, inequalities (11.2) and (11.4) are reversed.

Theorem 11.II. *The moments of all orders of a nonnegative random variable T whose survival function (complementary distribution function) is IFR exist.*

This property follows because $v(t)$ is bounded by an exponential for t sufficiently large (cf. (11.4)). We shall not present the details of the proof, the interest in this theorem being purely theoretical. Moreover, the existence of moments of all orders is assured under conditions considerably more general than the IFR condition. For example, it suffices that, for t sufficiently large, the failure rate $\lambda(t)$ be bounded below by a positive quantity [5, p. 43]. For IFR functions, we shall give below superior limits for the moments of T (Eq. (11.21) and Theorem 12.V).

Theorem 11.III. *If $v(t)$ is IFR and if one puts*[3]

$$v(t_k) = 1 - k\,, \quad \textit{where} \quad 0 < k < 1\,, \tag{11.11}$$

one has

$$v(t) \geqslant e^{-\lambda_0 t}\,, \qquad t \leqslant t_k\,, \tag{11.12}$$

$$v(t) \leqslant e^{-\lambda_0 t}\,, \qquad t \geqslant t_k\,, \tag{11.13}$$

where

$$\lambda_0 = -\frac{\ln(1-k)}{t_k}\,. \tag{11.14}$$

This theorem expresses the same property as Theorem 11.I in a different form, and it follows immediately from it. Indeed, we have

$$-\ln(1-k) = -\ln v(t_k) = \Lambda(t_k)\,,$$

thus

$$\lambda_0 = \Lambda(t_k)/t_k \geqslant \lambda(0)\,, \tag{11.15}$$

because of the convexity of the curve $\Lambda(t)$. In case (b) of Theorem 11.I we are given $\lambda_0 > \lambda(0)$. Here, we are given k and we deduce t_k by (11.11). If $\Lambda(t)$ is strictly convex for at least one value of t less than t_k, we have $\Lambda(t_k)/t_k > \lambda(0)$, and Theorem 11.I is applicable in particular for the value λ_0 given by (11.14), which is such that $\Lambda(t)$ and $\lambda_0 t$ are equal for $t = t_k$ (Fig. 11.3); inequalities (11.12) and (11.13) are then strictly satisfied for $t \neq t_k$ (Fig. 11.4). On the contrary, if the failure rate is constant for $t < t_k$, (11.12) is reduced to an equality.

[3] Thus t_k is a certain fractile of the distribution of T.

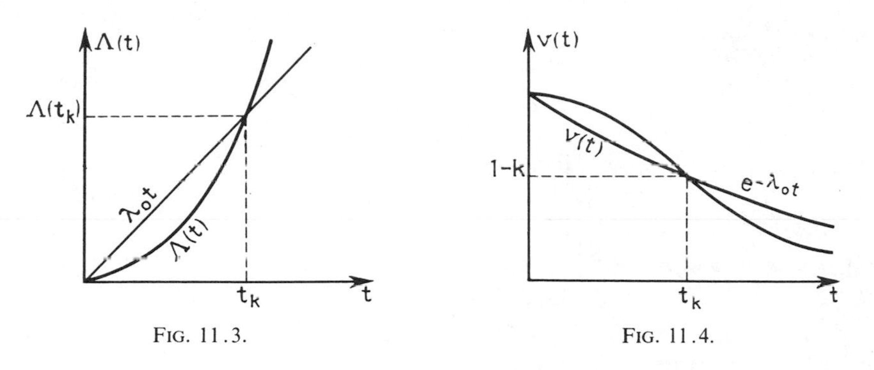

FIG. 11.3. FIG. 11.4.

Theorem 11.IV. *If $v(t)$ is IFR, has mean $\overline{T}$, and does not coincide with an exponential, one has*

$$v(t) > e^{-t/\overline{T}} \quad \text{for} \quad 0 < t < \overline{T} . \tag{11.16}$$

This theorem is again a particular application of Theorem 11.I, but it is also connected with properties of IFR functions relative to the mean lifetime $\overline{T}$.

We show first that

$$\overline{T} < 1/\lambda(0) \tag{11.17}$$

where $\lambda(0)$ is as before the failure rate at the origin, corresponding to the survival function $v(t)$. We have seen in (5.11) that

$$\overline{T} = \int_0^\infty v(u) \, du \tag{11.18}$$

or

$$\overline{T} = \int_0^\infty e^{-\Lambda(u)} \, du . \tag{11.19}$$

However, $\Lambda(t) \geqslant \lambda(0)t$, and the inequality is strict beyond a certain value of t if $v(t)$ is not an exponential law (see the proof of Theorem 11.I). It then follows that

$$e^{-\Lambda(t)} \leqslant e^{-\lambda(0).t} . \tag{11.20}$$

Since the inequality is strict for certain values of t, we may deduce

$$\int_0^\infty e^{-\Lambda(u)} \, du < \int_0^\infty e^{-\lambda(0) u} \, du = 1/\lambda(0)$$

and thus

$$\boxed{\bar{T} < 1/\lambda(0)} \; . \tag{11.21}$$

We apply Theorem 11.I for $\lambda_0 = 1/\bar{T}$, which is legitimate since, according to (11.21), $\lambda_0 > \lambda(0)$. Inequality (11.16) then results from (11.2), for $0 < t < t_0$, where t_0 is the abscissa of the point of intersection of the curves $\Lambda(t)$ and $\lambda_0 t = t/\bar{T}$ (Figs. 11.5 and 11.6). It thus suffices to prove that $\bar{T} \leqslant t_0$.

For this, we rely on Jensen's inequality, according to which, if $g(X)$ is a convex function of the random variable X, we have[4]

$$g(E[X]) \leqslant E[g(X)] \; . \tag{11.22}$$

Taking the lifetime T as the random variable X, and the function $\Lambda(\cdot)$ as the convex function $g(\cdot)$, the inequality (11.22) may be written as

$$\Lambda(\bar{T}) \leqslant E[\Lambda(T)] \; . \tag{11.23}$$

It is easy to see, however, that the random variable $\Lambda(T)$ follows an exponential law with mean 1. In fact, we have

$$\Pr\{\Lambda(T) > t\} = \Pr\{-\ln v(T) > t\} = \Pr\{v(T) < e^{-t}\} = e^{-t} .$$

The last inequality follows because the random variable $v(T)$ is uniformly

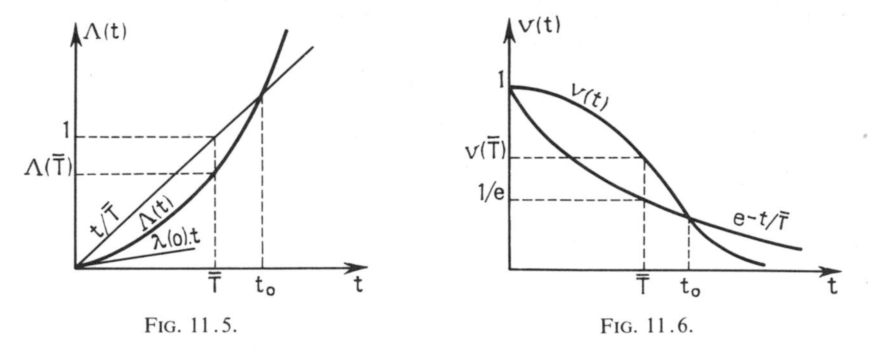

FIG. 11.5. FIG. 11.6.

[4] Inequality (11.22) represents, in the notations of the theory of probability, the continuous form of the following inequality: if $g(x)$ is a convex function in the interval (a, b), if $x_i \in (a, b)$, $i = 1, 2, \ldots, n$, and finally if $\lambda_i \geq 0$, $i = 1, 2, \ldots, n$, with $\sum_{i=1}^{n} \lambda_i = 1$, we have

$$g\left(\sum_{i=1}^{n} \lambda_i x_i\right) \leqslant \sum_{i=1}^{n} \lambda_i g(x_i) \; .$$

In (11.22), the role of the convex weightings $\lambda_1, \ldots, \lambda_n$ is played by the probability density of the random variable X.

distributed between 0 and 1[5]: The probability that $v(T) < e^{-t}$ is the same as the probability that $T > x$, where x is such that $v(x) = e^{-t}$; but, according to the same definition of $v(x)$, this probability is e^{-t}.

Thus we have

$$E[\Lambda(T)] = 1\,,$$

which could be verified easily by direct calculation.

Inequality (11.23) then gives

$$\Lambda(\overline{T}) \leqslant 1 \tag{11.24}$$

which may also be written as

$$-\ln v(\overline{T}) \leqslant 1$$

or again as

$$\boxed{v(\overline{T}) \geqslant 1/e}\,. \tag{11.25}$$

Relations (11.21) and (11.25) represent for IFR functions some interesting properties which we emphasize in the proof of Theorem 11.IV.

To return to this proof, relation (11.24), with the fact that for $t = \overline{T}$ the line $t/\overline{T}$ has an ordinate equal to 1 (Fig. 11.5), shows that the point of intersection of this line with the curve $\Lambda(t)$ has an abscissa at least equal to $\overline{T}$, which concludes the proof.

Theorem 11.V. *If $v(t)$ is IFR and has mean $\overline{T}$, one has*

$$v(t) \leqslant e^{-t\omega(t)} \quad \textit{for} \quad t > \overline{T}\,, \tag{11.26}$$

where $\omega(t)$ is the only positive solution of the equation

$$1 - \overline{T}\omega(t) = e^{-t\omega(t)}\,. \tag{11.27}$$

We note that if the inferior limit given for $v(t)$ in the interval $(0, \overline{T})$ by Theorem 11.IV is an exponential, the superior limit given above in the interval $(\overline{T}, \infty)$ is not an exponential function of t since the coefficient $\omega(t)$ varies with t.

The proof of this theorem will be given in Section 12 (Theorem 12.II). In Section 12 you will also find some other properties satisfied by IFR functions (Theorems 12.III–12.V).

[5] This property, which is general, is used in simulation in order to obtain an artificial sample of an arbitrary random variable from a sample of a uniform random variable.

12 Survival Functions with Increasing Failure Rate Averages

IFR survival functions have very interesting properties, but we shall see in Chapter IV that they are poorly suited to the study of complex equipment. Indeed, a system whose components have IFR survival functions does not necessarily itself have this same property.

Birnbaum and co-workers [9] have defined a class of survival functions that includes the class of IFR functions and that is stable with respect to the composition of structures to be studied in Section 26.

Definition. *Survival function with increasing failure rate average (IFRA). A survival function will be said to have an increasing failure rate average (IFRA) if and only if*

$$L(t) = \frac{\Lambda(t)}{t} \tag{12.1}$$

is nondecreasing.

Recall that the cumulative failure rate is defined by (4.20) as

$$\Lambda(t) = -\ln v(t) . \tag{12.2}$$

We have seen (Theorem 10.I) that for IFR laws of survival, the cumulative failure rate is convex; it then follows that the function $L(t)$ is nondecreasing, that is, that IFR functions are IFRA. More precisely, if for an IFR survival function there exists a value t_0 of t for which $\Lambda(t)$ is strictly convex (which is necessarily the case if the survival function being considered is not exponential), then the function $L(t)$ is strictly increasing for $t > t_0$.

Figure 12.1 represents the cumulative failure rate of an IFR function; for a point M with abscissa t arbitrarily placed on this curve, the slope of the line OM is $L(t)$, and it is clear that the convexity of $\Lambda(t)$ implies that this slope is nondecreasing. Figures 12.2 and 12.3 give two examples of IFRA functions that are not IFR.

Remark. In the case of a law of type II, where the instantaneous failure rate $\lambda(t)$ is defined everywhere, we have seen in (4.18) that

$$\Lambda(t) = \int_0^t \lambda(u)\, du . \tag{12.3}$$

Then

$$L(t) = \frac{1}{t}\int_0^t \lambda(u)\, du , \tag{12.4}$$

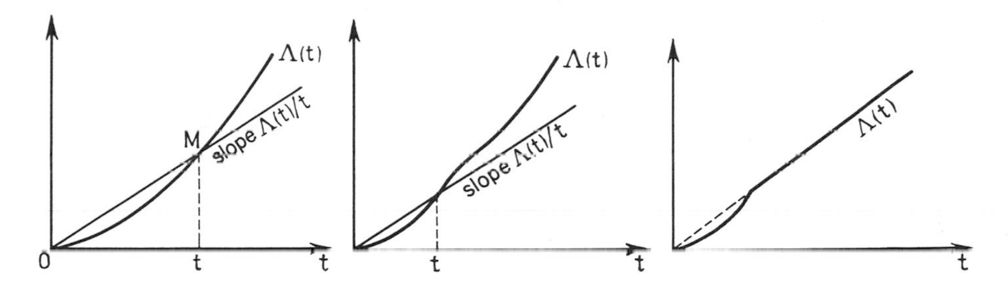

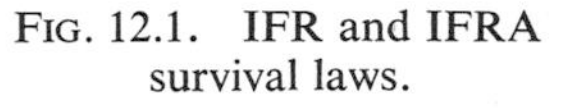

FIG. 12.1. IFR and IFRA survival laws.

FIG. 12.2 Survival law that is IFRA but not IFR.

FIG. 12.3. Survival law that is IFRA but not IFR.

that is, $L(t)$ is the mean of the instantaneous failure rate between 0 and t. Note, however, that $L(t)$ is not the failure rate in the interval $]0, t]$ as it was defined in (10.2); this last definition gives in fact

$$\mu(0\,;t) = 1 - v(t) \tag{12.5}$$

which may be written as

$$\mu(0\,;t) = 1 - e^{-\Lambda(t)}\,. \tag{12.6}$$

Properties of IFRA Functions. Some properties of IFR functions may be extended to IFRA functions in a slightly weakened form. We shall also indicate other properties that have not been mentioned in Section 11, but which are evidently valid for the more restricted class of IFR functions.

Theorem 12.I. *Let the survival function $v(t)$ be IFRA, $t_0 > 0$, and λ_0 be defined by*

$$\lambda_0 = -\frac{1}{t_0}\ln v(t_0) = L(t_0)\,. \tag{12.7}$$

Then one has

$$v(t) \geqslant e^{-\lambda_0 t}\,, \qquad 0 < t < t_0\,, \tag{12.8}$$

$$v(t_0) = e^{-\lambda_0 t_0}\,, \tag{12.9}$$

$$v(t) \leqslant e^{-\lambda_0 t}\,, \qquad t > t_0\,. \tag{12.10}$$

This theorem is identical, within the notations used, to Theorem 11.III, which used only the nondecreasing property of $L(t)$ (cf. (11.15)). In order to prove it, we first remark that λ_0 is defined by (12.7) in such a fashion that (12.9) is satisfied: λ_0 is the (constant) failure rate of the exponential function that passes through the point $(t_0, v(t_0))$. Similarly, we note that, according to (12.7), (12.2), and (12.1),

$$\lambda_0 = -\frac{\ln v(t_0)}{t_0} = \frac{\Lambda(t_0)}{t_0} = L(t_0)\,. \tag{12.11}$$

For $t < t_0$, one has, due to $L(t)$ being nondecreasing,

$$L(t) \leqslant L(t_0) \tag{12.12}$$

from which

$$\Lambda(t) \leqslant \lambda_0\, t \tag{12.13}$$

and, finally,

$$v(t) \geqslant \mathrm{e}^{-\lambda_0 t}\,. \tag{12.14}$$

For $t > t_0$, the inequalities are reversed.

The existence of moments of all orders of an IFRA survival function is assured for the same reasons as in the particular case of IFR functions (cf. Theorem 11.II).

On the contrary, Theorem 11.IV does not apply to the class of IFRA functions, as may be seen from a counterexample. Before that, let us point out that inequality (11.21) remains valid, at least in the nonstrict form

$$\overline{T} \leqslant 1/\lambda(0) \tag{12.15}$$

where $\lambda(0)$ is the failure rate at the origin. In order to prove inequality (11.25), we have used Jensen's inequality, and thus the convexity of the cumulative failure rate; we shall see that this inequality does not extend to IFRA functions.

We take as an example the following survival function (Fig. 12.4):

$$\begin{aligned} v(t) &= 1\,, & 0 \leqslant t < a \\ v(t) &= \mathrm{e}^{-\lambda_0 t}\,, & t \geqslant a\,. \end{aligned} \tag{12.16}$$

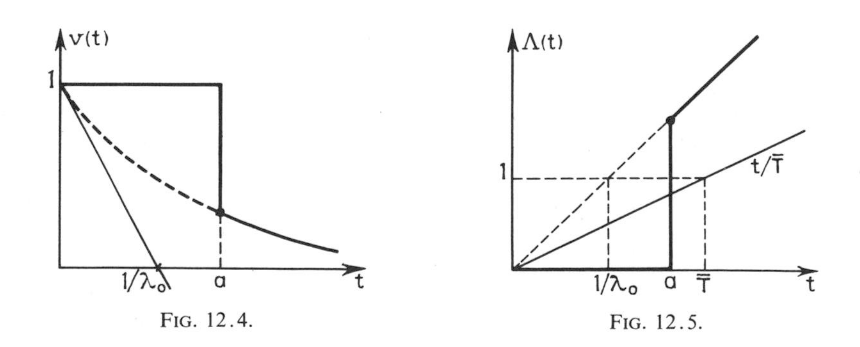

FIG. 12.4. FIG. 12.5.

The cumulative failure rate is given by

$$\Lambda(t) = -\ln v(t)\,, \tag{12.17}$$

or

$$\Lambda(t) = 0\,, \qquad 0 \leqslant t < a$$
$$\Lambda(t) = \lambda_0\, t\,, \qquad t \geqslant a \tag{12.18}$$

and the mean failure rate between 0 and t by

$$L(t) = 0\,, \qquad 0 \leqslant t < a$$
$$L(t) = \lambda_0\,, \qquad t \geqslant a\,. \tag{12.19}$$

It is indeed nondecreasing, and $v(t)$ is IFRA (but not IFR).

The mean lifetime is

$$\overline{T} = \int_0^\infty v(t)\,\mathrm{d}t = \int_0^a \mathrm{d}t + \int_a^\infty \mathrm{e}^{-\lambda_0 t}\,\mathrm{d}t = a + \frac{1}{\lambda_0}\,\mathrm{e}^{-\lambda_0 a}\,. \tag{12.20}$$

Note that

$$\overline{T} > a\,, \tag{12.21}$$

which shows that in order to obtain $\Lambda(t)$ or $v(\overline{T})$ it is necessary to use expressions valid for $t \geqslant a$. Thus

$$\Lambda(\overline{T}) = \lambda_0\,\overline{T} = \lambda_0\, a + \mathrm{e}^{-\lambda_0 a}\,. \tag{12.22}$$

If $a > 1/\lambda_0$, which is the case in Figs. 12.4 and 12.5, we have $\Lambda(\overline{T}) > 1$, from which $v(\overline{T}) < 1/e$, contradicting what is indicated by relations (11.24) and (11.25), which we have already proved for IFR functions. On the other hand, we have

$$\Lambda(t) > t/\overline{T} \quad \text{for} \quad t \geqslant a \tag{12.23}$$

from which

$$v(t) < \mathrm{e}^{-t/\overline{T}} \quad \text{for} \quad t \geqslant a\,. \tag{12.24}$$

These relations are valid in particular for $a \leqslant t < \overline{T}$, contrary to what is indicated by Theorem 11.IV.

Theorem 12.II. *If $v(t)$ is IFRA and has mean $\overline{T}$, one has*

$$v(t) \leqslant \mathrm{e}^{-t\omega(t)} \quad \textit{for} \quad t > \overline{T} \tag{12.25}$$

where $\omega(t)$ is the only positive solution to the equation

$$1 - \overline{T}\omega(t) = \mathrm{e}^{-t\omega(t)}\,. \tag{12.26}$$

This theorem recaptures Theorem 11.V for IFR functions, and the proof that we give here will a fortiori prove Theorem 11.V.

Let $\tau > \overline{T}$, and find $\omega(\tau) = \lambda_0 > 0$ such that relation (12.26) is satisfied:

$$1 - \lambda_0 \overline{T} = e^{-\lambda_0 \tau} \tag{12.27}$$

Thus

$$\frac{1 - e^{-\lambda_0 \tau}}{\lambda_0} = \overline{T}. \tag{12.28}$$

The first term may also be written as

$$\frac{1 - e^{-\lambda_0 \tau}}{\lambda_0} = \int_0^\tau e^{-\lambda_0 u} \, du. \tag{12.29}$$

Since $e^{-\lambda_0 u}$ is a decreasing function of λ_0, uniformly for $0 \leqslant u \leqslant \tau$, the integral in the second term is likewise a decreasing function of λ_0 that varies from τ (for $\lambda_0 = 0$) to 0 (for $\lambda_0 = \infty$); therefore there is a value of λ_0, and only one such value, for which the integral takes the value $\overline{T} < \tau$. This value is the solution of the equation

$$\int_0^\tau e^{-\lambda_0 u} \, du = \overline{T}. \tag{12.30}$$

Then consider the survival function (Fig. 12.7)

$$\begin{aligned} v_1(t) &= e^{-\lambda_0 t}, \qquad 0 \leqslant t < \tau, \\ &= 0, \qquad t \geqslant \tau, \end{aligned} \tag{12.31}$$

obtained from the exponential law with parameter λ_0 by introducing a limit of functioning τ (cf. Section 8). According to (12.30), this has a mean equal to $\overline{T}$. On the other hand, it is IFR (Fig. 12.6); we shall see that this is, among all IFRA survival functions with mean $\overline{T}$, the one that has the largest value for $t = \tau - \varepsilon$.

We show that there exists $t_0 < \tau$ such that

$$v(t) \leqslant e^{-\lambda_0 t} \quad \text{for} \quad t > t_0. \tag{12.32}$$

Relation

$$\int_0^\tau v(t) \, dt \leqslant \int_0^\infty v(t) \, dt = \overline{T} = \int_0^\tau e^{-\lambda_0 t} \, dt \tag{12.33}$$

shows that we may not have $v(t) > e^{-\lambda_0 t}$ for all t such that $0 < t < \tau$. There is thus a value t_0 of t such that $0 < t_0 < \tau$, and $v(t_0) \leqslant e^{-\lambda_0 t_0}$. Then, however, $\Lambda(t_0) \geqslant \lambda_0 t_0$, and since $\Lambda(t)/t$ is nondecreasing, we have $\Lambda(t) \geqslant \lambda_0 t$ for all $t > t_0$, from which (12.32) follows. In particular, for $t = \tau$, we have

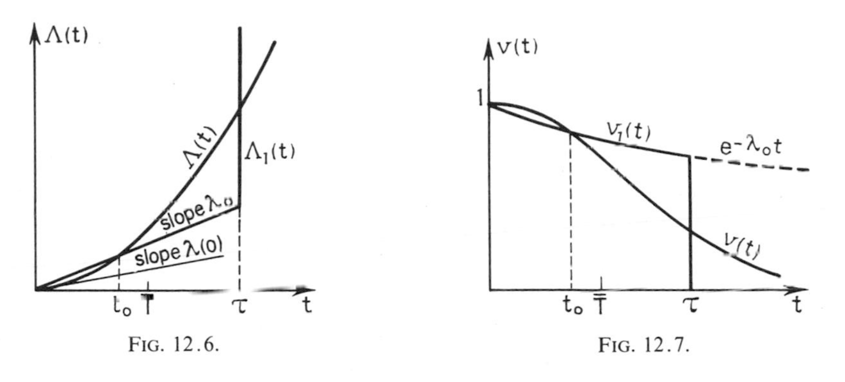

FIG. 12.6. FIG. 12.7.

$v(t) \leqslant e^{-\lambda_0 \tau}$. We then see that, for all values of τ greater than $\overline{T}$, the solution $\omega(\tau) = \lambda_0$ of Eq. (12.26) (or (12.30)) satisfies (12.25) for $t = \tau$.

We have seen above that the function $v_1(t)$ defined by (12.31) is IFR and has mean $\overline{T}$. Since, according to the theorem, *any* function that is IFRA with mean $\overline{T}$ satisfies the relation $v(\tau) \leqslant e^{-\lambda_0 \tau}$, and that, for $t = \tau - \varepsilon$, one has $v_1(t) = e^{-\lambda_0 t}$, it then follows that the superior limit fixed by the theorem may not be improved: the curve $e^{-t\omega(t)}$ is the superior envelope, for $t > \overline{T}$, of IFRA survival functions with mean $\overline{T}$.

Theorem 12.III. *If $v(t)$ is an IFRA survival function and has mean T, then there exists $\theta > 0$ such that*

$$v(t) \geqslant e^{-t/\overline{T}}, \qquad 0 < t < \theta, \tag{12.34}$$

$$v(t) \leqslant e^{-t/\overline{T}}, \qquad t > \theta. \tag{12.35}$$

We first remark that, if one has for a certain value $t_0 > 0$ of t,

$$v(t_0) > e^{-t_0/\overline{T}}, \tag{12.36}$$

it then follows that

$$\ln v(t_0) > -t_0/\overline{T}$$

and, according to (12.2),

$$\Lambda(t_0) < t_0/\overline{T} \tag{12.37}$$

or

$$L(t_0) = \frac{\Lambda(t_0)}{t_0} < \frac{1}{\overline{T}}.$$

Since $L(t)$ is nondecreasing, we shall then have

$$0 < t < t_0 \quad \Rightarrow \quad L(t) \leqslant L(t_0) < \frac{1}{\overline{T}} \quad \Leftrightarrow \quad v(t) > e^{-t/\overline{T}}. \tag{12.38}$$

In the same fashion, if

$$v(t_1) < e^{-t_1/\overline{T}}, \tag{12.39}$$

we shall have

$$t > t_1 \quad \Rightarrow \quad v(t) < e^{-t/\overline{T}}. \tag{12.40}$$

However, we have

$$\int_0^\infty v(t)\,dt = \overline{T} = \int_0^\infty e^{-t/\overline{T}}\,dt. \tag{12.41}$$

Since $v(t) \geqslant 0$ and $e^{-t/\overline{T}} \geqslant 0$, either $v(t) = e^{-t/\overline{T}}$ for all t, and the theorem is verified trivially for all θ, or there exists a t_0 satisfying (12.36) *and* a t_1 satisfying (12.39), evidently with $t_0 < t_1$. In the second case, the limits

$$t_m = \inf\{\, t/t > 0,\ v(t) < e^{-t/\overline{T}} \,\} \tag{12.42}$$

$$t_M = \sup\{\, t/v(t) > e^{-t/\overline{T}} \,\} \tag{12.43}$$

exist and satisfy the relation

$$t_m \leqslant t_M, \tag{12.44}$$

and the theorem is verified for all θ belonging to the closed interval $[t_m, t_M]$. Note that if $t_m < t_M$, we obtain $v(t) = e^{-t/\overline{T}}$ for all $t \in]t_m, t_M]$.

Theorem 12.IV. *If $v(t)$ is IFRA and has mean $\overline{T}$, and if $\varphi(t)$ is a nondecreasing function (respectively, nonincreasing), one has*

$$\int_0^\infty \varphi(t)\,v(t)\,dt \leqslant \int_0^\infty \varphi(t)\,e^{-t/\overline{T}}\,dt, \quad (\text{resp. } \geqslant). \tag{12.45}$$

As we have seen in Theorem 12.V, relation (12.45) permits us to compare the moments of a random variable whose survival function is IFRA to those of a random variable whose law is exponential.

Let θ be a positive number such that inequalities (12.34) and (12.35) of Theorem 12.III are satisfied. According to (12.41) we may write

$$\int_0^\infty \varphi(\theta)\,v(t)\,dt = \int_0^\infty \varphi(\theta)\,e^{-t/\overline{T}}\,dt = \varphi(\theta).\overline{T}. \tag{12.46}$$

Put

$$I = \int_0^\infty \varphi(t)\,v(t)\,dt - \int_0^\infty \varphi(t)\,e^{-t/\overline{T}}\,dt = \int_0^\infty \varphi(t)\,[v(t) - e^{-t/\overline{T}}]\,dt. \tag{12.47}$$

According to (12.46), we also have

$$I = \int_0^\infty [\varphi(t) - \varphi(\theta)]\,[v(t) - e^{-t/\overline{T}}]\,dt. \tag{12.48}$$

If, however, $\varphi(t)$ is nondecreasing, then

$$(12.49) \qquad 0 < t < \theta \quad \Rightarrow \quad \varphi(t) - \varphi(\theta) \leqslant 0 \quad \text{and} \quad v(t) - e^{-t/\overline{T}} \geqslant 0$$

and

$$(12.50) \qquad t > \theta \quad \Rightarrow \quad \varphi(t) - \varphi(\theta) \geqslant 0 \quad \text{and} \quad v(t) - e^{-t/\overline{T}} \leqslant 0 \, .$$

Thus

$$(12.51) \qquad I \leqslant 0 \, ,$$

and the theorem is proved in the case of a function $\varphi(t)$ that is nondecreasing. The proof is the same in the case of a nonincreasing function.

Theorem 12.V. *If $v(t)$ is IFRA and has mean $\overline{T}$, the moment of order r of the random variable T whose survival function is $v(t)$ satisfies the condition*

$$(12.52) \qquad E[T^r] \leqslant r\,!\,\overline{T}^r \, , \qquad r = 1, 2, 3, \ldots .$$

The proof is immediate from Theorem 12.IV. By definition we have

$$(12.53) \qquad E[T^r] = -\int_0^\infty t^r \, \mathrm{d}v(t) \, .$$

In the same fashion as in Section 5, an integration by parts permits us to write

$$(12.54) \qquad E[T^r] = r \int_0^\infty t^{r-1} \, v(t) \, \mathrm{d}t \, .$$

The function t^{r-1} being nondecreasing, Theorem 12.IV gives

$$(12.55) \qquad \int_0^\infty t^{r-1} \, v(t) \, \mathrm{d}t \leqslant \int_0^\infty t^{r-1} \, e^{-t/\overline{T}} \, \mathrm{d}t \, .$$

The second term of (12.55) represents (to within the factor r) the moment of order r of the exponential law, which we may easily calculate by making the change of variable $u = t/\overline{T}$:

$$(12.56) \qquad r \int_0^\infty t^{r-1} \, e^{-t/\overline{T}} \, \mathrm{d}t = r \int_0^\infty \overline{T}^r \, u^{r-1} \, e^{-u} \, \mathrm{d}u = r \overline{T}^r \, \Gamma(r) \, ,$$

where $\Gamma(r)$ is the Eulerian function of the second kind. We know that, for

$$r \in \{ 1, 2, 3, \ldots \} \, ,$$

we have $\Gamma(r) = (r - 1)!$. Relations (12.55) and (12.56) then give (12.52).

Using Theorem 12.V and inequality (12.15) together we obtain a bound on the moment of order r as a function of the failure rate at the origin $\lambda(0)$ of the survival law

$$(12.57) \qquad E(T^r) \leqslant \frac{r\,!}{[\lambda(0)]^r} \, .$$

Corollary. *The coefficient of variation* $(\sigma_T/\overline{T})$ *of an IFRA survival law is greater than or equal to 1.*

In fact, we have, according to (12.52),

$$E(T^2) \leqslant 2\,\overline{T}^2 \tag{12.58}$$

from which

$$\sigma_T^2 = E(T^2) - \overline{T}^2 \leqslant \overline{T}^2 . \tag{12.59}$$

Note that the superior limit is attained by the exponential law, for which $\sigma_T = \overline{T}$.

CHAPTER III

STUDY OF THE STRUCTURE OF SYSTEMS: STRUCTURE FUNCTIONS AND RELIABILITY NETWORKS

13 Introduction

In the first two chapters of this work we have considered a piece of equipment as a "black box" for which one may observe only whether or not it functions, and we have described probabilistic methods that allow the study of the lifetime of this black box. We now proceed to penetrate the box; most equipment is in fact quite complex, and the detailed study of its reliability supposes that it is considered as a "system" formed of simpler elements.

We shall proceed in two stages: In the present chapter, we shall see how the structure of a system may be described, that is, the way in which the state of a system depends on the state of its elements. This chapter does not use probabilistic concepts and may be considered independently of the first two.

In the following chapter, we shall combine the concepts that we then have at our disposal. We shall know how to represent and to study the lifetime of a system on one hand, and its elements on the other; and we shall know how to pass from the lifetimes of the elements to the lifetime of a system as a function of its structure.

14 Hypotheses on the Structure and Functioning of Systems

The theory developed in the sections that follow is applicable to systems or complex equipment[1a] that satisfy the following hypotheses:

(1) At any given instant, the equipment may be in only one of the following two states[1b]: it is functioning, or it is faulty.

(2) The equipment may be decomposed into r components, numbered as $1, \ldots, r$, in such a fashion that

each component is, at a given instant, either functioning or faulty;
the state of the equipment (good or failed) depends only on the state of its components.

We shall see that one may associate with such a piece of equipment a "structure function," or indeed, with certain reservations, a "reliability network." However, determining the algebraic expression of the structure function or constructing the reliability network associated with a given piece of equipment necessitates a deep understanding of these functions and networks; we shall dwell first on these more general and more abstract aspects.

In the following chapter, we shall pass by means of a very simple transformation from the structure function to the reliability function, and then to the survival function.

15 Structure Function

Consider a system S composed of r components e_i, $i = 1, 2, \ldots, r$. We first associate with each component e_i a state variable x_i such that

$$x_i = \begin{cases} 1 & \text{if the component } e_i \text{ is in a good state,} \\ 0 & \text{if the component } e_i \text{ is faulty.} \end{cases}$$

If $\mathbf{e} = \{ e_1, e_2, \ldots, e_r \}$ is the set of components, the r-tuple $(x_1, x_2, \ldots, x_r)$ will be called the "state of the set of components"; this will also be denoted by (x):

$$(x) = (x_1, x_2, \ldots, x_r)\,. \tag{15.1}$$

We know that there exist 2^r r-tuples such as (15.1); there are thus 2^r different states of the set of components.

[1a] One may speak of a system and the equipment that constitutes it, or of complex equipment and its components.

[1b] In Chapter VI we shall consider the case of equipment that has two types of failure.

Let y be the state variable of the system such that

$$y = \begin{cases} 1 & \text{if the system is in a good state,} \\ 0 & \text{if the system is faulty.} \end{cases}$$

Evidently, y depends on (x), and the hypotheses of Section 14 imply that there exists a function[2] $(x) \rightsquigarrow y$ which we denote by

$$y = \varphi(x) \quad \text{or} \quad y = \varphi(x_1, x_2, \ldots, x_r)\,. \tag{15.2}$$

This function will be called a "structure function" of the system. A structure function depending effectively (see Section 20) on r variables will be said to be of "order r."

Example 1 (components in series). Let

$$\begin{aligned} y &= \varphi(x_1, x_2, \ldots, x_r) \\ &= x_1 . x_2 . \ldots . x_r \\ &= \prod_{i=1}^{r} x_i\,. \end{aligned} \tag{15.3}$$

This structure function corresponds to a system that functions only under the condition that all its components are in a good state:

$$\begin{aligned} (\forall i,\ x_i = 1) &\Rightarrow (y = 1)\,, \\ (\exists i : x_i = 0) &\Rightarrow (y = 0)\,. \end{aligned} \tag{15.4}$$

Such a structure is said to be a "series structure"; one also says that the components are in series.

Example 2 (components in parallel). Let

$$\begin{aligned} y &= \varphi(x_1, x_2, \ldots, x_r) \\ &= 1 - (1 - x_1).(1 - x_2). \ldots .(1 - x_r) \\ &= 1 - \prod_{i=1}^{r} (1 - x_i)\,. \end{aligned} \tag{15.5}$$

This structure function corresponds to a system that functions only under the condition that at least one of its components is in a good state:

$$\begin{aligned} (\exists i : x_i = 1) &\Rightarrow (y = 1)\,, \\ (\forall i,\ x_i = 0) &\Rightarrow (y = 0)\,. \end{aligned} \tag{15.6}$$

Such a structure is said to be a "parallel structure"; we also say that the components are in parallel. The use of two or more components in parallel

[2] This will signify that to any r-tuple $(x_1, x_2, \ldots, x_r)$ corresponds one and only one value of y.

when only one would suffice introduces a redundance in the equipment. We shall return to this point later (Chapter V).

Remark. The two examples given above indeed form a structure function since, for any state of the set of components (r-tuple formed of 0's and 1's), the function y may take only the value 0 or 1. The operation defined by (15.3) and that defined by (15.5) may lead only to the realization of functions that are structure functions whatever the number of repetitions of these operations; we thus obtain the set of monotone structures (see Section 21).

Example 1. Let

$$y = \varphi(x_1, x_2, x_3) = x_1[1 - (1 - x_2)(1 - x_3)] . \tag{15.7}$$

This structure function is obtained by applying operation (15.5) to x_2 and x_3:

$$\varphi_1(x_2, x_3) = 1 - (1 - x_2)(1 - x_3); \tag{15.8}$$

then by applying operation (15.3) to x_1 and $\varphi_1(x_2, x_3)$:

$$y = x_1 \varphi_1(x_2, x_3) = x_1[1 - (1 - x_2)(1 - x_3)] . \tag{15.9}$$

x_1	x_2	x_3	$\varphi(x_1, x_2, x_3)$
0	0	0	0
0	0	1	0
0	1	0	0
0	1	1	0
1	0	0	0
1	0	1	1
1	1	0	1
1	1	1	1

FIG. 15.1.

One may easily verify that $\varphi(x_1, x_2, x_3)$ takes only the values 0 and 1; in fact, one obtains the results given in Fig. 15.1. Indeed $\varphi(x_1, x_2, x_3)$ is a structure function.

Example 2.

$$y = \varphi(x_1) = 1 - x_1 . \tag{15.10}$$

This very simple function cannot be obtained using only operations (15.3) or (15.5); it is, however, a structure function since it assumes only the value 0 or 1. It is a nonmonotone structure function (see the remarks at the end of Sections 16 and 21). The same holds for the following two examples.

Example 3.

$$y = \varphi(x_1, x_2, x_3) = (1 - x_1)(1 - x_2)(1 - x_3) . \quad (15.11)$$

One may easily verify that y assumes only the value 0 or 1.

Example 4.

$$y = 1 - [1 - (1 - x_1)(1 - x_2) x_3][1 - x_1(1 - x_2)(1 - x_3)] ; \quad (15.12)$$

this function is also a structure function.

Comparison of r-tuples. Suppose that the elements E_i of set $E = \{E_1, E_2, \ldots, E_k\}$ are ordered; we shall use the notation $E_i \preccurlyeq E_j$ to indicate the existence of this order relation between the elements of the pair (E_i, E_j).

Now consider two r-tuples

$$(a) = (a_1, a_2, \ldots, a_r) \quad \text{and} \quad (b) = (b_1, b_2, \ldots, b_r) . \quad (15.13)$$

We shall then say that:

(A) (a) equals (b), denoted $(a) = (b)$, if and only if

$$\forall i , \quad i = 1, 2, \ldots, r : a_i = b_i . \quad (15.14)$$

(B) (a) exceeds (b), denoted $(a) \succcurlyeq (b)$ (we may also say "dominates"), if and only if

$$\forall i , \quad i = 1, 2, \ldots, r : a_i \geqslant b_i . \quad (15.15)$$

(C) (a) is less than (b), denoted $(a) \preccurlyeq (b)$ (we may also say "is dominated by"), if and only if

$$\forall i , \quad i = 1, 2, \ldots, r : a_i \leqslant b_i . \quad (15.16)$$

(D) (a) strictly dominates (b), denoted $(a) \succ (b)$, if and only if

$$a_i \geqslant b_i \quad \text{and} \quad \exists i : a_i > b_i , \quad (15.17)$$

(E) (a) is strictly dominated by (b), denoted $(a) \prec (b)$, if and only if

$$\forall i : a_i \leqslant b_i \quad \text{and} \quad \exists i : a_i < b_i . \quad (15.18)$$

Examples.

$$(1, 0, 0, 1, 0) \prec (1, 1, 0, 1, 0) \prec (1, 1, 0, 1, 1) . \quad (15.19)$$

(15.20) $(1, 0, 1, 0, 1)$ may not be compared with $(1, 1, 0, 0, 1)$.

Boolean Lattice of States of the Set of Components. The order relation defined above introduces in the set of 2^r r-tuples, that is, in the set of states of the set of components, a Boolean lattice structure. Figure 15.2 gives a representation (Hasse diagram) of the lattice corresponding to a system of three components. A 3-tuple (*a*) is dominated by (*b*) if and only if one may climb from (*a*) to (*b*) along the connections of the diagram. For example, (001) is dominated by (011), (101), and (111), but not by (000), (010), or (110).

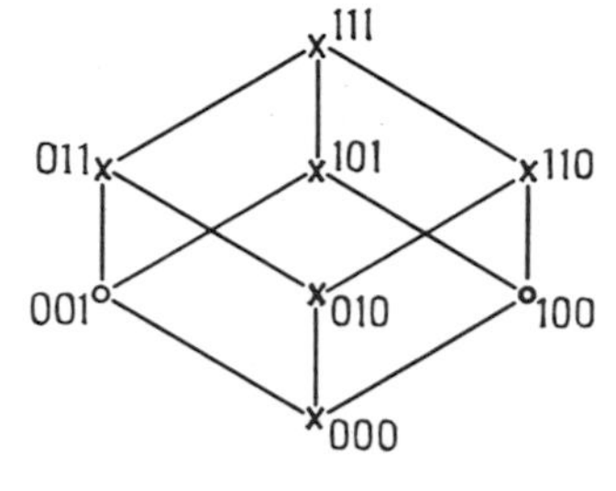

FIG. 15.2.

In addition, one may, as we have done in Fig. 15.2, use this same representation to indicate the values taken by the structure function, that is, the state of the system for each of the states of the set of components. It suffices to use a different symbol according as the system functions ($\varphi = 1$) or fails to function ($\varphi = 0$). Figure 15.2 corresponds to the structure function (15.12); the states in which the system functions are represented by small circles, and those in which it does not function by crosses.

Graphical Representation of a System. It is convenient to represent with a nonoriented graph[3] a system of components being studied with a view toward examining its reliability. In Figs. 15.3 and 15.4 are presented some systems with components in series and in parallel. Figure 15.5 gives a more complex example.

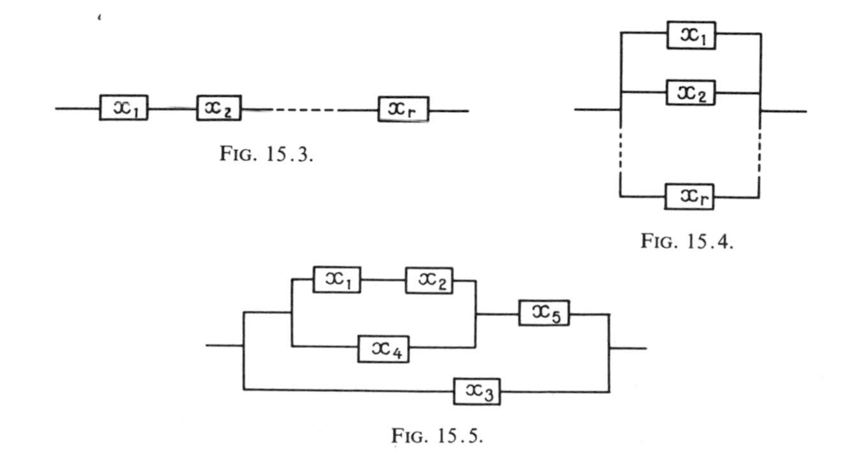

FIG. 15.3.

FIG. 15.4.

FIG. 15.5.

[3] See the review of the theory of graphs in Section 18.

The representation used is not in fact a physical reality, except in particular cases (certain electrical, mechanical, or pneumatic assemblages, for example). It is used here only with a pedagogic goal; the introduction in Section 19 of reliability networks will give it a more precise sense.

Subsets of Components. Let $\mathbf{e} = \{ e_1, e_2, \ldots, e_r \}$ be the set of components of a system; any part of $\mathbf{e}$ will constitute a subset of components.

Example. Let

(15.21) $$\mathbf{e} = \{ e_1, e_2, e_3, e_4, e_5 \};$$

then

(15.22) $$\mathbf{e}_1 = \{ e_2, e_4, e_5 \}$$

will be a subset of components, for example, as will

(15.23) $$\mathbf{e}_2 = \{ e_1, e_2 \} .$$

For algebraic convenience, one defines an empty subset of $\mathbf{e}$, designated by $\varnothing$ as usual. One also defines the complement of a subset as

(15.24) $$\complement_{\mathbf{e}}\, \mathbf{e}_1 = \mathbf{e} - \mathbf{e}_1 ;$$

when there is no question of ambiguity in the nature of the reference set, this complement is denoted $\bar{\mathbf{e}}_1$.

Examples. Considering (15.22):

(15.25) $$\bar{\mathbf{e}}_1 = \{ e_1, e_3 \} .$$

Considering (15.23):

(15.26) $$\bar{\mathbf{e}}_2 = \{ e_3, e_4, e_5 \} .$$

Cardinality of a Subset. The cardinality of a subset refers to the number of components constituting this subset.

16 Links and Cuts

Links of a Structure.[4] Consider a structure defined by the function

(16.1) $$y = \varphi(x_1, x_2, \ldots, x_r)$$

[4] The term *path* is also frequently used, but we prefer here to use *link* to avoid any confusion with the notion of path in graph theory, a notion that is close to that used here but not identical, as we shall see in Section 18.

and whose components are

$$\mathbf{e} = \{ e_1, e_2, \ldots, e_r \} . \tag{16.2}$$

Let

$$\mathbf{a} = \{ e_i, i \in I \} \subset \mathbf{e} \tag{16.3}$$

be a subset of components defined by the subset $I \subset \{ 1, 2, \ldots, r \}$ of indices. If

$$\left.\begin{array}{l} x_i = 1 \, , i \in I \\ x_i = 0 \, , i \notin I \end{array}\right\} \Rightarrow \quad y = 1 \, , \tag{16.4}$$

one says that $\mathbf{a}$ is a *link*. A link is thus a subset of components such that if all the components of this subset are in a good state, and if the other components have failed, then the system is in a good state.

Example. Consider the structure function

$$y = x_3 + x_4\, x_5 + x_1\, x_2\, x_5 - x_3\, x_4\, x_5 - x_1\, x_2\, x_3\, x_5 - x_1\, x_2\, x_4\, x_5 + x_1\, x_2\, x_3\, x_4\, x_5 \, . \tag{16.5}$$

One may verify that this is a structure function (in fact, it is that corresponding to the structure given in Fig. 15.5).

The subset of components

$$\mathbf{a} = \{ e_1, e_2, e_3, e_5 \} = \{ e_i, i \in I \} \quad \text{with} \quad I = \{ 1, 2, 3, 5 \} \tag{16.6}$$

is a link. In effect, put $x_1 = x_2 = x_3 = x_5 = 1$ and $x_4 = 0$ in (16.5); we have

$$\varphi(1, 1, 1, 0, 1) = 1 \, . \tag{16.7}$$

Similarly, the subset $\mathbf{a}' = \{ e_3 \}$ is a link.

On the contrary,

$$\mathbf{a}'' = \{ e_1, e_2, e_4 \} \tag{16.8}$$

is not a link. We have

$$\varphi(1, 1, 0, 1, 0) = 0 \, . \tag{16.9}$$

Cut of a Structure. Consider again the structure defined by (16.1) and (16.2). Let

$$\mathbf{b} = \{ e_j, j \in J \} \subset \mathbf{e} \tag{16.10}$$

be the subset of components whose indices belong to J. If

$$\left.\begin{array}{l} x_j = 0 \, , j \in J \\ x_j = 1 \, , j \notin J \end{array}\right\} \Rightarrow \quad y = 0 \, , \tag{16.11}$$

one says that **b** is a *cut*: If the components of **b** have failed, and if the other components are in a good state, then the system does not function.

Example. Consider again the structure function (16.5). The subset of components

$$\mathbf{b} = \{\, e_2, e_3, e_5 \,\} \tag{16.12}$$

is a cut. In effect, put $x_2 = x_3 = x_5 = 0$ and $x_1 = x_4 = 1$ in (16.5); we have

$$\varphi(1, 0, 0, 1, 0) = 0 + 0 + 0 - 0 - 0 - 0 + 0 = 0\,. \tag{16.13}$$

On the other hand,

$$\mathbf{b}' = \{\, e_1, e_2, e_4 \,\} \tag{16.14}$$

is not a cut. We have

$$\varphi(0, 0, 1, 0, 1) = 1\,. \tag{16.15}$$

Minimal Link. If a link

$$\mathbf{a} = \{\, e_{a_1}, e_{a_2}, \ldots, e_{a_l} \,\} \tag{16.16}$$

is such that there does not exist a subset $\mathbf{a}' \subset \subset \mathbf{a}$ that is also a link, we say that **a** is a "minimal link."

Example. Once again take the structure function (16.5); considering the subset

$$\mathbf{a} = \{\, e_4, e_5 \,\} \tag{16.17}$$

we have

$$\varphi(0, 0, 0, 1, 1) = 1\,. \tag{16.18}$$

Thus this subset is a link.

One may verify that $\{\, e_4 \,\}$ and $\{\, e_5 \,\}$ are not links. It then follows that $\{\, e_4, e_5 \,\}$ is a minimal link. Moreover, this structure function possesses three minimal links, which are

$$\{\, e_1, e_2, e_5 \,\},\ \{\, e_4, e_5 \,\},\ \{\, e_3 \,\}\,.$$

Minimal Cut. If a cut

$$\mathbf{b} = \{\, e_j, j \in J \,\} \tag{16.19}$$

is such that there does not exist a subset $\mathbf{b}' \subset \subset \mathbf{b}$ that is also a cut, we say that **b** is a "minimal cut."

Example. Again looking at the structure function (16.5), we have that

$$\mathbf{b} = \{ e_1, e_3, e_4 \}; \tag{16.20}$$

is a cut since

$$\varphi(0, 1, 0, 0, 1) = 0 + 0 + 0 - 0 - 0 - 0 + 0 = 0. \tag{16.21}$$

One may verify that the three subsets $\{ e_1, e_3 \}$, $\{ e_1, e_4 \}$, and $\{ e_3, e_4 \}$ are not cuts, and neither are $\{ e_1 \}$, $\{ e_3 \}$, and $\{ e_4 \}$.

The structure function has in fact three minimal cuts, which are

$$\{ e_1, e_3, e_4 \}, \{ e_2, e_3, e_4 \}, \{ e_3, e_5 \}.$$

Complementary Subset of a Link or a Cut. It follows immediately from the definitions of links and cuts that one has the following properties:

$$\mathbf{a} \text{ is a link} \quad \Leftrightarrow \quad \bar{\mathbf{a}} \text{ is not a cut}; \tag{16.22}$$

$$\mathbf{b} \text{ is a cut} \quad \Leftrightarrow \quad \bar{\mathbf{b}} \text{ is not a link}. \tag{16.23}$$

To prove these logical identities, let us remark that "$\mathbf{a}$ is a link and $\bar{\mathbf{a}}$ is a cut" or, similarly, "$\mathbf{b}$ is a cut and $\bar{\mathbf{b}}$ is a link" are contradictory statements. For example, "$\mathbf{a}$ is a link" and "$\bar{\mathbf{a}}$ is a cut" both imply the same state of the set of components, but different values of the state of the system.

Examples. Referring once again to the structure (16.5), one may easily verify that

$$\{ e_4, e_5 \} \text{ is a link} \quad \text{and} \quad \{ e_1, e_2, e_3 \} \text{ is not a cut};$$

$$\{ e_2, e_3, e_5 \} \text{ is a cut} \quad \text{and} \quad \{ e_1, e_4 \} \text{ is not a link}.$$

Remark. The definitions and properties that we have given may very easily be interpreted in Fig. 15.5, which visualizes the system whose structure function is given by (16.5). There are two other properties that are evident in this figure, or any other analogous figure:

$$\mathbf{a} \text{ is a link}, \quad \mathbf{a} \subset \mathbf{a}' \quad \Rightarrow \quad \mathbf{a}' \text{ is a link}; \tag{16.24}$$

$$\mathbf{b} \text{ is a cut}, \quad \mathbf{b} \subset \mathbf{b}' \quad \Rightarrow \quad \mathbf{b}' \text{ is a cut}. \tag{16.25}$$

We shall see that these properties are verified for the structures that are said to be monotone (Section 21), which are representable by a reliability network (Section 19), but structure functions exist for which properties (16.24) and (16.25) are not satisfied.

Example. For the structure function (15.12), one may verify that $\{ e_3 \}$ is a link, but that $\{ e_2, e_3 \}$ is not one; $\{ e_1 \}$ is a cut, but $\{ e_1, e_2 \}$ is not (moreover, this is the complement of the link $\{ e_3 \}$).

17 Mathematical Properties of Links and Cuts. Duality

Theorem 17.I. *Any r-tuple* $(x_1, x_2, \ldots, x_r)$ *with* $x_i = 0$ *or* 1 *corresponds either to a link or to a cut.*

Indeed, let $\mathbf{e}_1$ be the subset of components for which $x_i = 1$. If $\varphi(x_1, \ldots, x_r) = 1$, $\mathbf{e}_1$ is a link, and $\bar{\mathbf{e}}_1$ is not a cut; if $\varphi(x_1, \ldots, x_r) = 0$, $\bar{\mathbf{e}}_1$ is a cut, and $\mathbf{e}_1$ is not a link.

On a lattice of states of a set of components, an example of which was given in Fig. 15.2, each vertex (or node) of the lattice represents an r-tuple, to which corresponds a link (vertex where the system functions, represented by a small circle), or to a cut (vertex where the system does not function, represented by a cross).

In this same representation of a lattice, the properties indicated in Section 16 may be expressed in the following fashion:

To the relation $x^{(1)} \preccurlyeq x^{(2)}$ between r-tuples corresponds the inclusion relation $\mathbf{e}_1 \subset \mathbf{e}_2$ between the subsets defined as above, that is, the subsets for which $x_i = 1$ in the r-tuple considered. A minimal link then corresponds to a lattice vertex represented by a small circle and connected downward only with crosses (for example, (001) represents the minimal link $\{ e_3 \}$ in Fig. 15.2). A minimal cut corresponds to a cross that is connected only to circles through upward connections (in Fig. 15.2 the only minimal cut is the empty set $\varnothing$, corresponding to the vertex 111). Properties (16.22) and (16.23) simply express the fact that at a given vertex of a lattice, the system may not both function and not function at the same time. Finally, the fact that properties (16.24) and (16.25) are not satisfied by the structure function (15.12), to which corresponds Fig. 15.2, is translated by the fact that not all the vertices to which one may climb up from a circle are circles (001 is a circle, 011 is not), and not all the vertices to which one may climb down from a cross are crosses (take the same example, considered in the reverse sense).

Theorem 17.II. *Let* A_k *be the number of links having k components and* B_k *the number of cuts having k components. Then*

$$A_k + B_{r-k} = \binom{r}{k} \tag{17.1}$$

where $\binom{r}{k}$ *is the number of combinations of r objects taken k at a time.*

As a consequence, we associate with each of the $\binom{r}{k}$ subsets of k components, the r-tuple obtained by taking the value 1 for the components belonging to the subset, and the value 0 for the others. According to the proof of Theorem 17.I, there then corresponds to each subset either a link having k components, or a cut having $r - k$ components.

Length of a System. The length of a system S is the smallest integer λ such that $A_\lambda > 0$. In other words, this is the number of components in the (minimal) link having the smallest cardinality. This length is denoted by $\lambda(S)$.

Example. The length of the system corresponding to (16.5) is equal to 1 since the link having the smallest number of components is $\{ x_3 \}$.

Width of a System. The width of a system S is the smallest integer μ such that $B_\mu > 0$. In other words, this is the number of components of the (minimal) cut having the smallest cardinality. This width is denoted by $\mu(S)$.

Example. The width of the system corresponding to (16.5) is equal to 2 (cut $\{ x_3, x_5 \}$).

Theorem 17.III. *The length $\lambda(S)$ and the width $\mu(S)$ of the same system S satisfy the inequality*

$$\lambda(S) + \mu(S) \leqslant r + 1 \,, \tag{17.2}$$

where r is the number of components of the system.

PROOF. A set containing $\lambda(S) - 1$ components cannot be a link, and therefore its complement, which has $r - \lambda(S) + 1$ components, is necessarily a cut. By virtue of the definition of the width $\mu(S)$ of a system, we obtain

$$\mu(S) \leqslant r - \lambda(S) + 1 \,,$$

which proves (17.2).

Degenerate Systems. One says that a system is degenerate if and only if

$$\varphi(x) \equiv 1 \tag{17.3}$$

or

$$\varphi(x) \equiv 0 \,. \tag{17.4}$$

In other words, whatever the state of its set of components, the system is always functioning or always broken down. From another point of view, one would have either a zero length or a zero width, if the notions of length and width were extended to zero values.

Duality. Let S be a system whose structure function is $\varphi(x)$; the system $\bar{S}$ whose structure function is

$$\bar{\varphi}(x) = 1 - \varphi(1 - x) \tag{17.5}$$

where $(1 - x) = (1 - x_1, 1 - x_2, \ldots, 1 - x_r)$, is called the "dual of S."

Example. Consider the structure function

$$y = \varphi(x_1, x_2, x_3, x_4) = x_4 + x_1 x_3 + x_2 x_3 - x_1 x_2 x_3 - x_1 x_3 x_4 - x_2 x_3 x_4 + x_1 x_2 x_3 x_4 . \tag{17.6}$$

This function may be rewritten in the form

$$y = 1 - (1 - [1 - (1 - x_1)(1 - x_2)]\, x_3)(1 - x_4) \tag{17.7}$$

from which it more readily appears that the corresponding system has a representation as in Fig. 17.1.

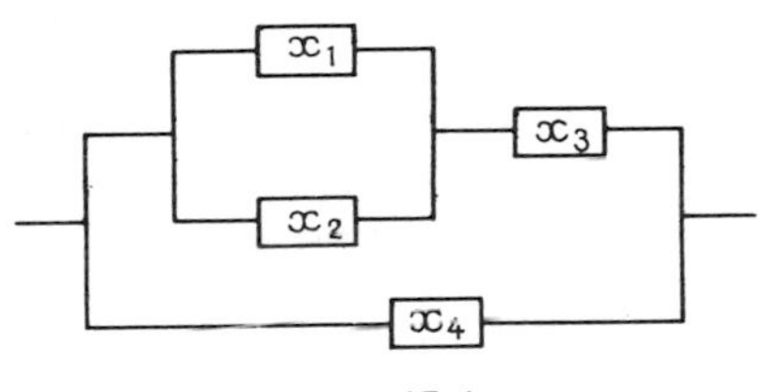

FIG. 17.1.

Now compute

$$\overline{\varphi}(x) = 1 - \varphi(1 - x) . \tag{17.8}$$

Instead of replacing x_i by $1 - x_i$ in (17.6), we do this in (17.7), which is easier;

$$\bar{y} = \overline{\varphi}(x) = 1 - \varphi(1 - x) = [1 - (1 - x_1 x_2)(1 - x_3)]\, x_4 = x_3 x_4 + x_1 x_2 x_4 - x_1 x_2 x_3 x_4 . \tag{17.9}$$

The representation corresponding to (17.9) is given in Fig. 17.2.

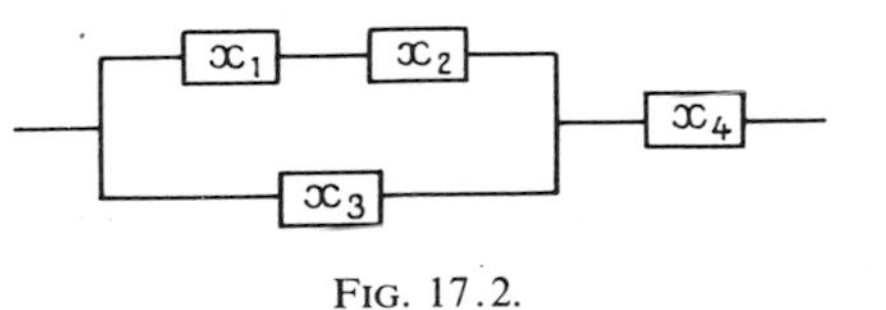

FIG. 17.2.

Properties of Duality. We give several properties which the reader may easily prove:

(1) $\overline{[\overline{\varphi}(x)]} = \varphi(x)$, $\overline{(\overline{S})} = S$.
(2) A link of S is a cut of $\overline{S}$ and conversely.
(3) A cut of S is a link of $\overline{S}$ and conversely.

(4) A minimal link of S is a minimal cut of $\bar{S}$ and conversely.
(5) A minimal cut of S is a minimal link of $\bar{S}$ and conversely.

18 Review of the Theory of Graphs[5]

Graphs. Consider a finite set **S** and the product set **S** × **S**. Let **U** be a subset of **S** × **S**; the ordered pair

$$G = (\mathbf{S}, \mathbf{U})$$

is called a graph. The elements of **S** are called "vertices of the graph"; the elements of **U**, which are pairs of vertices, are called "arcs of the graph."

Figure 18.1a gives a graphical representation of a graph. The vertices are represented by points, and the arcs by continuous lines joining the vertices and bearing an arrowhead. The set of vertices is

$$\mathbf{S} = \{ A, B, C, D \} \tag{18.1}$$

and the set of arcs is

$$\mathbf{U} = \{ (AB), (AD), (BB), (BC), (BD), (CC), (DA), (DB), (DC), (DD) \} . \tag{18.2}$$

Given an arc such as (A, B), A is called "the initial end of the arc" and B is its "terminal end." An arc such as (B, B) is called a "loop." A vertex B is said to be a "successor" of A if (A, B) is an arc; for example, in the graph of Fig. 18.1a, C is not a successor of A.

A graph may also be described by its Boolean matrix, that is, by a square matrix whose rows and columns correspond to the vertices of the graph and whose elements are valued 0 or 1 according as the pair corresponding to the vertices does or does not belong to **U** (see Fig. 18.1b).

In order to store a graph in the memory of a computer, one usually uses a "dictionary of successors" and/or a "dictionary of precedents." As the name indicates, the dictionary of successors gives for each vertex of the graph its list of successors, which amounts to describing the Boolean matrix row by row (Fig. 18.1c). The dictionary of precedents, on the other hand, gives for each vertex X the list of vertices that are initial points of an arc having the vertex X for a terminal point; it corresponds to the columns of the Boolean matrix.

[5] For more details, the reader should consult the following works: C. Berge, *Théorie des Graphes et ses Applications*. Dunod, Paris, 1958; A. Kaufmann, *Graphs, Dynamic Programming, and Finite Games*. Academic Press, New York, 1967; A. Kaufmann, *Introduction à la Combinatorique*. Dunod, Paris, 1968; B. Roy, *Algèbre Moderne et Théorie des Graphes*, 2 Vols. Dunod, Paris, 1969–1970; R. G. Busacker and T. L. Saaty, *Finite Graphs and Networks*. McGraw-Hill, 1965.

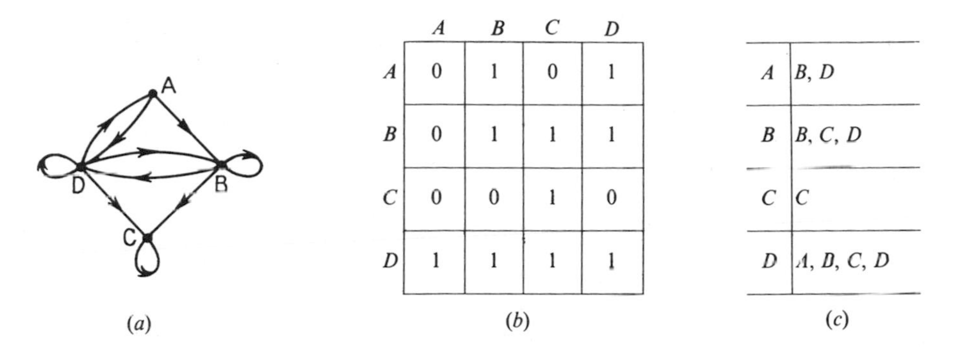

	A	B	C	D
A	0	1	0	1
B	0	1	1	1
C	0	0	1	0
D	1	1	1	1

(b)

A	B, D
B	B, C, D
C	C
D	A, B, C, D

(c)

FIG. 18.1. (a) Graphical representation; (b) Boolean matrix; (c) dictionary of successors.

***r*-Fold Graphs.** In certain applications it is convenient to allow that there exists between certain vertices, not one arc (at most), but two or more (Fig. 18.2).

The corresponding concept is no longer a graph in the sense of the definition above; we shall call this an "r-fold graph," where r is the maximal number of arcs having the same initial point and the same terminal point. Such arcs will have to be distinguished, for example, by an index, as we have done in Fig. 18.2.

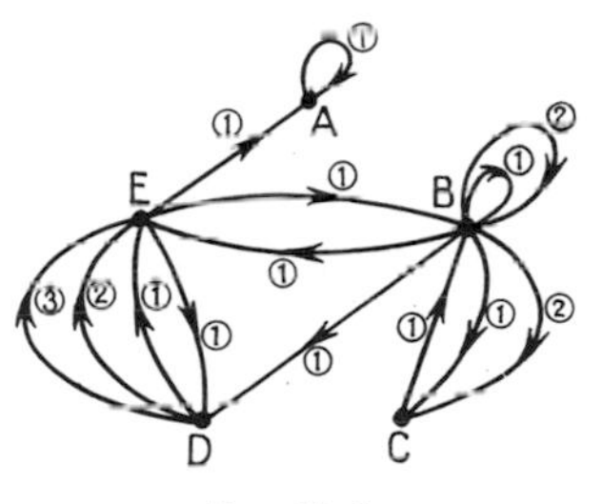

FIG. 18.2.

In a more precise fashion, one may define an r-fold graph as formed of

(a) a set **S** of vertices, and

(b) a set **U** of arcs and a mapping Ω of **U** into **S** × **S** that associates to each arc an initial point and a terminal point.

The notation that we shall use to designate the arcs will in fact dispense with explicit mention of the mapping Ω: for example, $(DE)_2$ evidently designates an arc with initial point D and terminal point E. Figure 18.2 thus represents a 3-fold graph

$$G = (\mathbf{S}, \mathbf{U}) \tag{18.3}$$

with

(18.4) $$\mathbf{S} = \{ A, B, C, D, E \}$$

and

(18.5) $$\mathbf{U} = \{ (AA)_1, (BB)_1, (BB)_2, (BC)_1, (BC)_2, (BD)_1, (BE)_1, (CB)_1, (DE)_1, (DE)_2, (DE)_3, (EA)_1, (EB)_1, (ED)_1 \} .$$

Other Definitions. We now proceed to define on r-fold graphs several concepts that we shall have use for in this work. These concepts are evidently applicable to graphs as we have initially defined them, which may be considered as 1-fold graphs. In the following parts of this book we shall use the term "graph" to designate r-fold graphs, whatever the value of r.

Given a graph $G = (\mathbf{S}, \mathbf{U})$, a partial graph of G is a graph $G_p = (\mathbf{S}, \mathbf{U}')$ such that $\mathbf{U}' \subset \mathbf{U}$. Thus one may obtain a partial graph of G by suppressing certain arcs. Then the graph in Fig. 18.3 is a partial graph of the graph presented in Fig. 18.2.

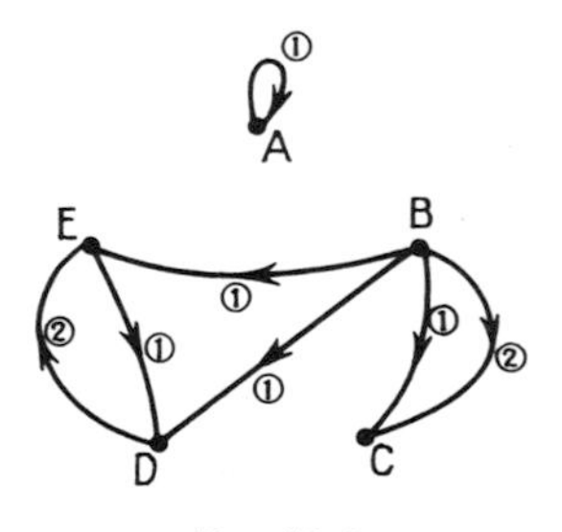

FIG. 18.3.

A path is a sequence of arcs

(18.6) $$\mu = (u_1, u_2, \ldots, u_l) \quad \text{with} \quad u_i \in \mathbf{U}\,, \quad i = 1, \ldots, l,$$

such that the terminal point of each arc u_i $(i = 1, 2, \ldots, l-1)$ coincides with the initial point of the following arc u_{i+1}. A path such that the terminal point of the arc u_l coincides with the initial point of the arc u_1 is called a "circuit." Examples: In the graph of Fig. 18.2, $((BD)_1, (DE)_2, (EA)_1, (AA)_1)$ is a path; $((BC)_2, (CB)_1, (BE)_1, (ED)_1, (DE)_3, (EB)_1)$ is a circuit.

The length of a path is the number of arcs that it contains.

A path is said to be "elementary" if it does not pass twice through the same vertex. The path and the circuit given as examples above are not elementary; on the other hand, $((BD)_1, (DE)_2, (EA)_1)$ is elementary.

Given two arbitrary vertices S_i and S_j of a graph G, if there exists a path μ from S_i to S_j, there exists then an elementary path from S_i to S_j; in the preceding example, the path $((BD)_1, (DE)_2, (EA)_1, (AA)_1)$ passes through the vertex A twice; it thus contains a circuit, here reduced to the loop $(AA)_1$; by suppressing all the circuits contained in the path μ whose existence is known, it is rendered elementary, which proves the existence of such an elementary path. We shall use this property in Section 23.

Consider a subset of vertices $\mathbf{S}_1 \subset \mathbf{S}$. A cut of the graph G relative to the subset $\mathbf{S}_1$ will be the set $\omega^-(\mathbf{S}_1)$ of arcs whose initial ends do not belong to S_1 and whose terminal ends belong to $\mathbf{S}_1$. For example, if we put $\mathbf{S}_1 = \{A, E\}$ in the graph of Fig. 18.2, the corresponding cut is

$$\omega^-(\{A, E\}) = \{(BE)_1, (DE)_1, (DE)_2, (DE)_3\}. \tag{18.7}$$

The interest in this notion and the term used may be explained in the following fashion. In the example above, suppose that one is interested in the paths running from a vertex not in S_1, such as C, to a vertex of S_1, such as A. One may easily convince oneself that any path from C to A contains at least one arc of the cut (18.7). It follows that in the partial graph obtained by suppressing all the arcs of the cut, there is no path from C to A, moreover none from any vertex of $\mathbf{S} - \mathbf{S}_1$ whatever to any vertex of $\mathbf{S}_1$.

Before ending, we introduce one last notion: A tree with root A will be a graph without circuits such that:

(1) A is a vertex that is not the terminal end of any arc, and
(2) each vertex other than A is the terminal end of one arc.

Note that a tree is a 1-fold graph. For this reason it is not necessary to provide indices for the arcs in the graph of Fig. 18.4.

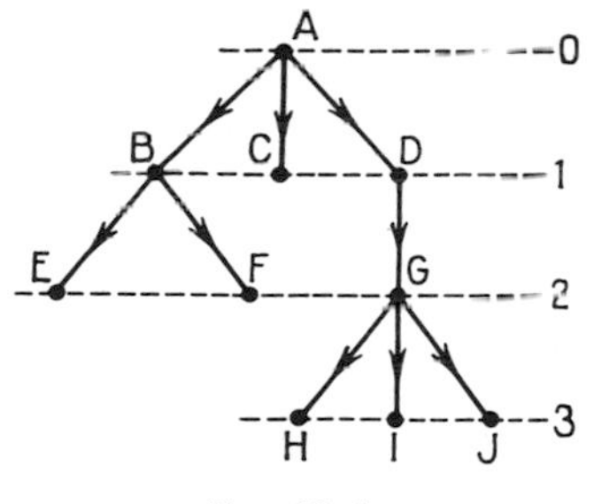

FIG. 18.4.

The vertices of a tree may be divided into levels: If A represents level 0, any other vertex will have level equal to the length of the (unique) path going from A to this vertex.

19 Reliability Networks

A reliability network $\mathfrak{R}$ defined on a set $\mathbf{e} = \{e_1, e_2, \ldots, e_r\}$ of components consists of:

(1) an r-fold graph $G = (\mathbf{S}, \mathbf{U})$ without loops and in which two vertices $O \in \mathbf{S}$ and $Z \in \mathbf{S}$ are distinguished and called, respectively, the origin and end, and

(2) a mapping $\Delta: \mathbf{U} \overset{\Delta}{\leadsto} \mathbf{e}$ such that

$$\Omega(u_j) = (S_i, S_k), \quad \Omega(u_{j'}) = (S_i, S_k) \Rightarrow \Delta(u_j) \neq \Delta(u_{j'}),$$

where Ω is the mapping that corresponds to each arc the pair of its ends (cf. Section 18).

The mapping Δ corresponds a component to each arc of the graph. Several arcs may correspond to the same component,[6] and it may happen that there is no arc corresponding to a given component. Figure 19.1 gives an example of a reliability network where

(19.1) $\mathbf{e} = \{ e_1, e_2, e_3, e_4 \}$,

(19.2) $\mathbf{S} = \{ O, Z, A, B, C \}$,

(19.3) $\mathbf{U} = \{ (O, A)_2, (O, A)_3, (O, B), (A, B), (A, Z), (B, C), (B, Z), (C, B), (Z, B) \}$.

The mapping Δ is indicated by the e_i attached to the arcs.

To simplify the figures, if two symmetric arcs (X_i, X_j) and (X_j, X_i) concern the same component, it will be convenient to replace these two symmetric arcs with a single arc with two opposing arrowheads (Fig. 19.2). To further simplify the presentation, the names of the vertices may be omitted when this will introduce no confusion.

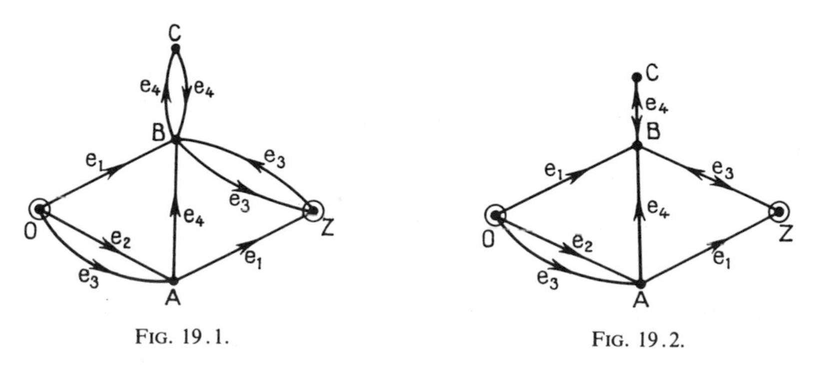

FIG. 19.1. FIG. 19.2.

Links of a Reliability Network. To any subset $\mathbf{e}_1 \subset \mathbf{e}$ of components one may correspond the partial graph $G_p(\mathbf{e}_1)$ of the graph G, obtained by

[6] This is subject to the condition that they do not have the same end points. Arcs having the same end points, like the two arcs from O to A in Fig. 19.1, will be marked with the index of the component to which it corresponds through the mapping Δ.

retaining only the arcs of G to which correspond a component belonging to $\mathbf{e}_1$:

$$G_p(\mathbf{e}_1) = (\mathbf{S}, \mathbf{U}_p(\mathbf{e}_1)) \tag{19.4}$$

with

$$\mathbf{U}_p(\mathbf{e}_1) = \{ u \in \mathbf{U} \mid \Delta(u) \in \mathbf{e}_1 \} . \tag{19.5}$$

A link of a network $\mathfrak{R}$ will refer to a subset $\mathbf{a} \subset \mathbf{e}$ of components such that there exists in the graph $G_p(\mathbf{a})$ a path from O to Z.

Example. $\{ e_1, e_2, e_3 \}$ is a link of the network in Fig. 19.2, as may be seen in Fig. 19.3 which represents the corresponding partial graph. To each link there corresponds one or more paths of the graph. Inversely, to a path $\mu = (u_1, u_2, \ldots, u_l)$ from O to Z there corresponds a link $\mathbf{a}$ formed by the images of the arcs $u_1, u_2, \ldots, u_l$ through the mapping Δ.

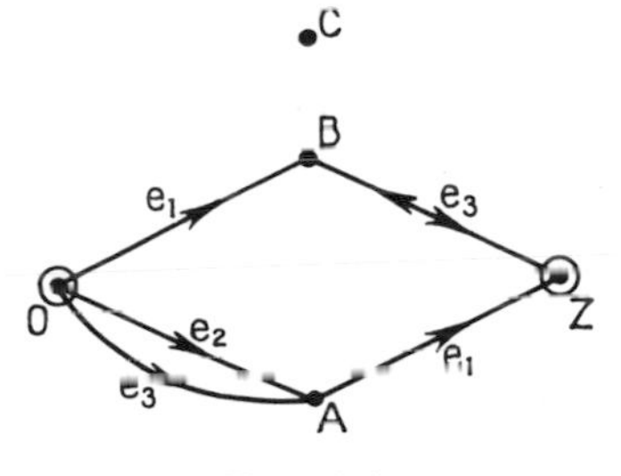

FIG. 19.3.

Example. To the path $\mu = ((OA)_3, AB, BZ)$ there corresponds the link $\mathbf{a} = \{ e_3, e_4 \}$.

Cuts of a Reliability Network. A cut of a reliability network $\mathfrak{R}$ will refer to a subset $\mathbf{b} \subset \mathbf{e}$ of components such that the subset of arcs $\mathbf{U}_p(\mathbf{b})$ defined by (19.5) contains a cut of the graph G relative to a subset of vertices including O but excluding Z.

Example. $\{ e_1, e_2, e_3 \}$ is a cut of the network (and at the same time a link); the complementary subset $\{ e_4 \}$ is neither a cut nor a link.

To any cut $\mathbf{b}$ of a network there thus corresponds one or more cuts of the graph, included in $\mathbf{U}_p(\mathbf{b})$. In the example above, the set of arcs $\{ AZ, BZ \}$ is a cut of the graph that does not include the arcs OB, $(OA)_2$, and $(OA)_3$.

Note that the concepts of cut in a graph and of cut in a reliability network are not identical. It would no doubt be preferable to use two different terms as we have done in the case of paths and links,[7] but we have not found a satisfactory equivalent for the descriptive term "cut."

System Defined by a Reliability Network. Given a reliability network $\mathfrak{R}$, consider the system S having **e** for its set of components and such that for each state of the set of components in which the components of the subset $\mathbf{e}_1$ are in a good state and those of the complementary subset $\overline{\mathbf{e}_1}$ have failed, the system functions if $\mathbf{e}_1$ is a link of the reliability network $\mathfrak{R}$, and it fails in the contrary case. This system S is defined without ambiguity from the network $\mathfrak{R}$. Its structure function[8] $\varphi(x_1, x_2, \ldots, x_r)$ takes the value 1 for any state $x = (x_1, x_2, \ldots, x_r)$ such that the subset of components for which $x_i = 1$ is a link of the network; this then is also a link of the structure function. If the subset of components is not a link of the network, φ takes the value 0, and the subset is not a link of the structure function. It is easy to see, similarly, that the network and the structure function have the same cuts.

Example. Figure 19.4 represents the table of values of the structure function φ of the system defined by the network of Fig. 19.2. For the state (1, 1, 1, 0) of the set of components, one has $\varphi = 1$ since $\{e_1, e_2, e_3\}$ is a link; on the other hand, $\varphi(0, 0, 1, 0) = 0$ since $\{e_3\}$ is not a link. One may equivalently reason in terms of cuts: $\varphi(0, 0, 0, 1) = 0$ since $\{e_1, e_2, e_3\}$ is a cut. (Note

x_1	x_2	x_3	x_4	φ
0	0	0	0	0
0	0	0	1	0
0	0	1	0	0
0	0	1	1	1
0	1	0	0	0
0	1	0	1	0
0	1	1	0	0
0	1	1	1	1
1	0	0	0	0
1	0	0	1	0
1	0	1	0	1
1	0	1	1	1
1	1	0	0	1
1	1	0	1	1
1	1	1	0	1
1	1	1	1	1

FIG. 19.4.

[7] Hansel [28] uses the term *path* to designate what we have called a link; then to a link of the network corresponds a partial graph of the graph.

[8] See Section 15.

that it is the subset of components for which $x_i = 0$ that must be considered.) On the other hand, $\varphi(0, 0, 1, 1) = 1$ since $\{ e_1, e_2 \}$ is not a cut.

The identity between the links and cuts of a reliability network and those of the structure function defined by this network permits one to apply to reliability networks the notions defined in Sections 16 and 17. We shall now briefly review these notions using the language of reliability networks.

Minimal Cuts and Minimal Links. A link **a** is minimal if no subset $\mathbf{a}' \subset\subset \mathbf{a}$ is a link of the network. A cut **b** is minimal if no subset $\mathbf{b}' \subset\subset \mathbf{b}$ is a cut of the network.

Examples. In the network of Fig. 19.2, $\{ e_1, e_2, e_3 \}$ is not a minimal link; $\{ e_1, e_3 \}$ is one and is also a minimal cut. $\{ e_1, e_4 \}$ is another example of a minimal cut.

Complementarity Relations. Recall properties (16.22) and (16.23):

(19.6) $\qquad$ **a** is a link $\Leftrightarrow$ $\bar{\mathbf{a}}$ is not a cut,

(19.7) $\qquad$ **b** is a cut $\Leftrightarrow$ $\bar{\mathbf{b}}$ is not a link.

Degenerate Networks. A network is degenerate if:

(1) it possesses no link (the system never functions), or

(2) it possesses no cut (the system functions whatever the state of its components); the extremities O and Z are then identical.

Example (Fig. 19.5) The networks $\mathcal{R}_1$ (Fig. 19.5a) and $\mathcal{R}_2$ (Fig. 19.5b) are degenerate.

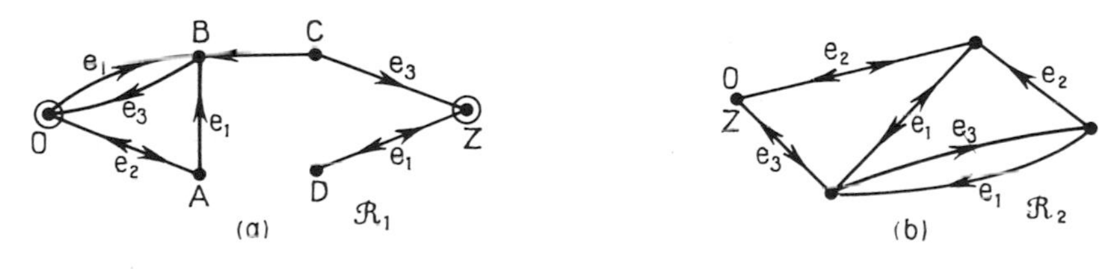

FIG. 19.5.

Fundamental Property of Reliability Networks. Adding one or more arc to a graph cannot suppress a path existing between the origin O and the end Z. The repair of a broken component thus could not entail the failure of a system representable by a reliability network (this hypothesis is reasonable but not absolute for real systems).

This property may be written in the following form:

$$\mathbf{a}' \subset \mathbf{a}, \quad (\mathbf{a}' \text{ is a link}) \quad \Rightarrow \quad (\mathbf{a} \text{ is a link}). \tag{19.8}$$

One has the same property for cuts:

$$\mathbf{b}' \subset \mathbf{b}, \quad (\mathbf{b}' \text{ is a cut}) \quad \Rightarrow \quad (\mathbf{b} \text{ is a cut}). \tag{19.9}$$

In effect, by adding arcs to a cut in a graph, one still obtains a cut. Properties (19.8) and (19.9) may be expressed in the following theorem.

Theorem 19.I. *In a reliability network, a subset of components including a link is also a link; a subset of components including a cut is also a cut.*

In more concrete terms, one may say that the systems defined by reliability networks are such that repair of a broken component cannot entail failure of the system, and that the failure of a component cannot entail the functioning of a failed system. We have already remarked at the end of Section 16 that properties (16.24) and (16.25) are not satisfied by all structure functions. We shall return to this point in Section 21.

Theorem 19.II. *A cut contains at least one component from each link, and a link contains at least one component from each cut. In other words, any link and any cut have at least one component in common.*

In fact, let **a** be a link and **b** be a cut; suppose they have no component in common, that is, $\mathbf{a} \subset \bar{\mathbf{b}}$. Then, according to the preceding theorem, $\bar{\mathbf{b}}$ is a link, and according to the property of complementarity (19.6), **b** is not a cut, which is contradictory to the hypotheses.

Theorem 19.III. *Any link includes at least one minimal link, and any cut includes at least one minimal cut.*

In fact, let **a** be a link; we seek a subset of **a** that is also a link. Two cases are possible:

(1) No subset of **a** is a link; then **a** is a minimal link, and the theorem is verified.

(2) A subset $\mathbf{a}_1 \subset \mathbf{a}$ is a link; we then begin again and seek a subset of $\mathbf{a}_1$ that is a link, and so on. With **a** having a finite number of components, we necessarily end with a minimal link, which may be empty if the network is degenerate.

Similar reasoning shows that any cut includes a minimal cut.

Theorem 19.IV. *In a nondegenerate network, a set* **b** *of components is a cut if and only if it contains at least one component of each minimal link.*

Let **b** be a subset containing a component of each minimal link; $\bar{\mathbf{b}}$ includes no minimal link, and is thus not a link (Theorem 19.III), and thus **b** is a cut (property (19.7)). Conversely, if **b** is a cut, $\bar{\mathbf{b}}$ is not a link, and can include no minimal link.

Theorem 19.V. *In a nondegenerate network, a set* **a** *of components is a link if and only if it contains a component from each minimal cut.*

(Same proof as for the preceding theorem.)

Length and Width of a Network. We shall preserve for reliability networks the terminology used in Section 17 for structure functions in Theorems 17.II and 17.III. Thus the numbers A_k, B_k, λ, and μ signify

A_k : number of links having k components
B_k : number of cuts having k components,
λ : length of a network:

$$(k < \lambda) \;\Rightarrow\; (A_k = 0)\,, \tag{19.10}$$

μ : width of a network:

$$(k < \mu) \;\Rightarrow\; (B_k = 0)\,. \tag{19.11}$$

Theorems 17.II and 17.III then stand valid for reliability networks. Furthermore, one may state the following theorem, proved in 1956 by Moore and Shannon [40].

Theorem 19.VI. *The number r of components of a network is at least equal to the product of the length and width of this network*

$$r \geqslant \lambda . \mu\,. \tag{19.12}$$

We do not reproduce here the somewhat long proof proposed by Moore and Shannon.

Typical Networks[9]

"Series" Networks. Each component is a cut. The set of components is the only link.

[9] The notions series, parallel, and bridge are common in the theory of electrical networks; a sketch of a reliability network is topologically analogous to that of an electrical network, but the meanings are very different.

Example (Fig. 19.6). One would still have a series network if one or more of these arcs had not its symmetrical arc in the direction of Z toward O.

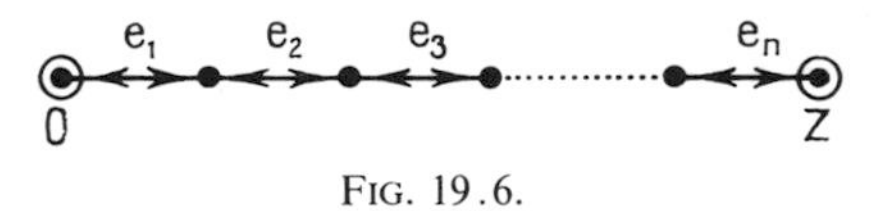

FIG. 19.6.

"*Parallel*" *Networks.* Each component is a link. The set of components is the only cut.

Example (Fig. 19.7). One would still have a parallel network if one or more of these arcs had not its symmetrical arc in the direction of Z toward O.

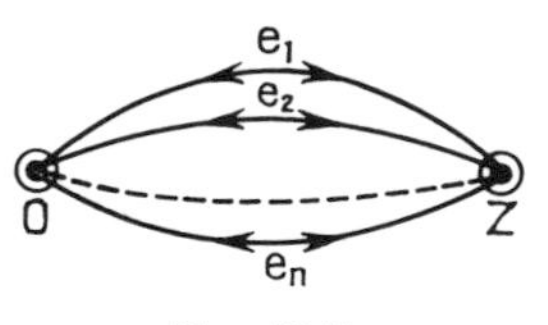

FIG. 19.7.

"*Bridge*" *Networks.* Such networks have the configuration presented in Figs. 19.8 and 19.9.

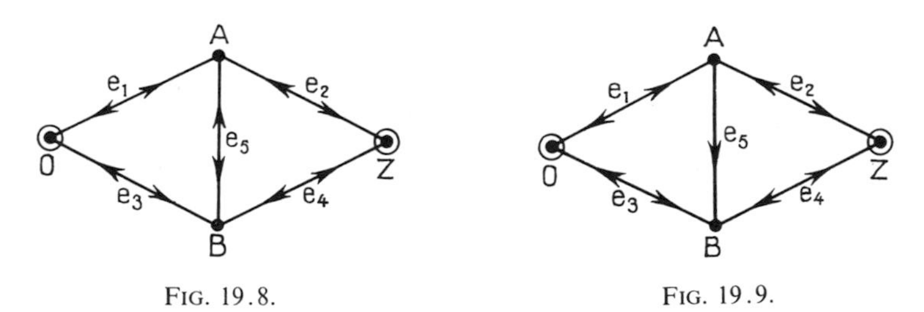

FIG. 19.8. FIG. 19.9.

The network of Fig. 19.8 possesses four minimal links and four minimal cuts, which are, respectively,

(19.13) $\{e_1, e_2\}$, $\{e_1, e_4, e_5\}$, $\{e_3, e_4\}$, $\{e_2, e_3, e_5\}$,

and

(19.14) $\{e_1, e_3\}$, $\{e_1, e_4, e_5\}$, $\{e_2, e_4\}$, $\{e_2, e_3, e_5\}$.

A variant (Fig. 19.9) possesses three minimal links and four maximal cuts; these are, respectively,

(19.15) $\{e_1, e_2\}$, $\{e_1, e_4, e_5\}$, $\{e_3, e_4\}$

and $\{e_1, e_3\}$, $\{e_1, e_4\}$, $\{e_2, e_4\}$, $\{e_2, e_3, e_5\}$.

Other variants are possible with fewer components than five. Thus the bridge of Fig. 19.10 has three minimal links and four minimal cuts.

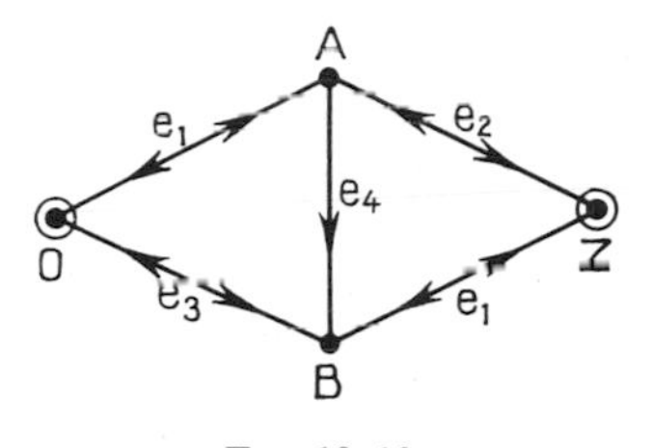

FIG. 19.10.

"Parallel–Series" Networks. Such networks have the configuration presented in Fig. 19.11. The components joining two adjacent vertices in this network constitute a cut. One obtains a link by taking a component in each cut.

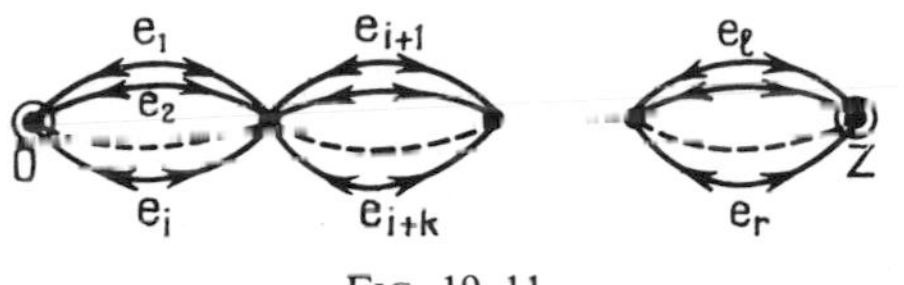

FIG. 19.11.

"Series–Parallel" Networks. These are networks having a configuration like that indicated in Fig. 19.12.

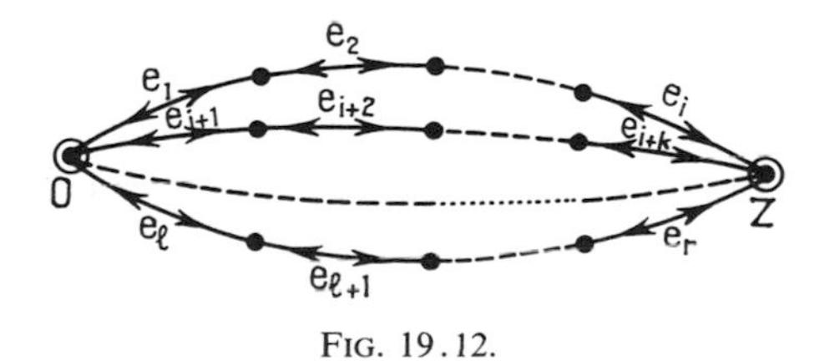

FIG. 19.12.

Duality in Reliability Networks. A network $\mathfrak{R}$ is dual to a network $\mathfrak{R}$ if it has the following property: Any link of $\mathfrak{R}$ is a cut of $\mathfrak{R}$ and vice versa. With the aid of Theorems 19.IV and 19.V one may easily prove that any cut of $\mathfrak{R}$ is a link of $\mathfrak{R}$.

Properties of duality in reliability networks.

(1) $\overline{(\overline{\mathfrak{R}})} = \mathfrak{R}$.
(2) A link of $\mathfrak{R}$ is a cut of $\overline{\mathfrak{R}}$ and vice versa.
(3) A cut of $\mathfrak{R}$ is a link of $\overline{\mathfrak{R}}$ and vice versa.
(4) A minimal link of $\mathfrak{R}$ is a minimal cut of $\overline{\mathfrak{R}}$ and vice versa.
(5) A minimal cut of $\mathfrak{R}$ is a minimal link of $\overline{\mathfrak{R}}$ and vice versa.

20 Equivalence between Structure Functions and Reliability Networks

Equivalent Structure Functions. To any system satisfying the hypotheses set forth in Section 14 one may associate a unique structure function, in the sense that for each state of the set of components, a unique value can indeed be defined. This structure function may however be expressed by various algebraic formulas.

In a slightly more general fashion, we consider:

(1) three disjoint sets of components $\mathbf{e}$, $\mathbf{e}_1$, $\mathbf{e}_2$, containing, respectively, r, r_1, and r_2 components; we designate by $x = (x_1, x_2, \ldots, x_r)$, $u = (u_1, u_2, \ldots, u_{r_1})$, and $v = (v_1, v_2, \ldots, v_{r_2})$ the states of these sets of components;

(2) two structure functions $\varphi_1(x, u)$ and $\varphi_2(x, v)$, defined, respectively, on the sets of components $\mathbf{e} \cup \mathbf{e}_1$ and $\mathbf{e} \cup \mathbf{e}_2$.

We shall say that the structure functions $\varphi_1(x, u)$ and $\varphi_2(x, v)$ are "equivalent" if

$$\forall x \in \{0, 1\}^r, \quad u \in \{0, 1\}^{r_1}, \quad v \in \{0, 1\}^{r_2} : \varphi_1(x, u) = \varphi_2(x, v).$$

It follows immediately from this definition that, whatever the state x of the set of components $\mathbf{e}$, $\varphi_1(x, u)$ has a value independent of u, and $\varphi_2(x, v)$ has a value independent of v. The components of $\mathbf{e}_1$ and of $\mathbf{e}_2$ are called "useless components"; their state, good or bad, has no influence on the functioning of the systems represented by the equivalent structure functions $\varphi_1(x, u)$ and $\varphi_2(x, v)$.

It may not appear to be very realistic to consider systems having useless components. In fact, this is necessary in certain cases because systems are often called upon to play various roles or to have various modes of use. A separate analysis of reliability must be made for each of various uses, and certain components may be useless for some of these.

The possible existence of useless components necessitates several precautions for comparison of structure functions. Thus it follows from Theorem 17.I that a system is completely defined by the list of its links (or by the list of cuts): for any r-tuple $a = (a_1, a_2, \ldots, a_r)$, if $\mathbf{a}$ is the subset of components for which $a_i = 1$, one has $\varphi(a_1, a_2, \ldots, a_r) = 1$ when $\mathbf{a}$ occurs in the list of links and $\varphi = 0$ in the contrary case. This is true only if there is no ambiguity in the set of components over which one considers the system to be defined. In order to avoid any difficulties, one may use the following theorem.

Theorem 20.I. *Two structure functions, not necessarily defined on identical sets of components, are equivalent if, after suppression of any possible useless components, every link of one is a link of the other (or indeed, any cut of one is a cut of the other).*

Example. The following four structure functions are equivalent:

(20.1) $$\varphi_1(x_1, x_2) = [1 - (1 - x_1)(1 - x_2)]\, x_1^2,$$

(20.2) $$\varphi_2(x_1, x_2) = [1 - (1 - x_1)(1 - x_2)]\, x_1,$$

(20.3) $$\varphi_3(x_1, x_2) = 1 - (1 - x_1)(1 - x_1 x_2),$$

(20.4) $$\varphi_4(x_1) = x_1 .$$

One may verify that this is so either by expanding the expressions above and noting that $x_i = x_i^2 = x_i^3 = \cdots$, since x_i takes only the values 0 and 1, or by establishing the list of links (or of cuts). Figure 20.1 gives this list, in the first part before suppression of the useless component x_2 for φ_1, φ_2, and φ_3, and in the second after suppression of the useless component.

	φ_1	φ_2	φ_3	φ_4
Links before suppression of useless component	$\{e_1\}$ $\{e_1, e_2\}$	$\{e_1\}$ $\{e_1, e_2\}$	$\{e_1\}$ $\{e_1, e_2\}$	$\{e_1\}$ $\{e_1, e_2\}$
Links after suppression of useless component	$\{e_1\}$	$\{e_1\}$	$\{e_1\}$	$\{e_1\}$

FIG. 20.1.

Equivalent Reliability Networks. The preceding definition for equivalent structure functions may be extended to reliability networks. Two reliability networks $\mathcal{R}_1$ and $\mathcal{R}_2$ are said to be "equivalent," and we write

(20.5) $$\mathcal{R}_1 = \mathcal{R}_2$$

if, after suppression of possible useless components (see the definition below), any link of $\mathcal{R}_1$ is a link of $\mathcal{R}_2$ and vice versa (respectively, any cut of $\mathcal{R}_1$ is a cut of $\mathcal{R}_2$ and vice versa).

Definition. *A component x_i of a reliability network is useless if every link containing x_i remains a link after suppression of x_i (respectively, if every cut containing x_i remains a cut after suppression of x_i).*

Example (Fig. 20.2). $\{e_1, e_2, e_3\}$ is the only link containing e_2; $\{e_1, e_3\}$ is also a link; therefore e_2 is useless. On the contrary, $\{e_1, e_2\}$ is not a link, and therefore e_3 is not useless.

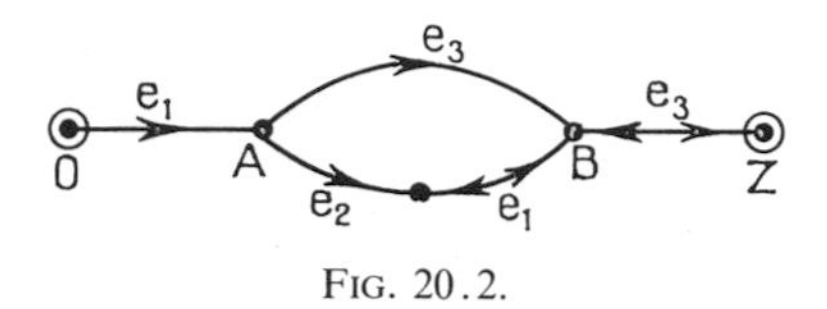

FIG. 20.2.

Another Example (Fig. 20.3). We show that e_1 is useless. The links that contain e_1 are

(20.6) $$\{e_1, e_2\},$$

(20.7) $$\{e_1, e_2, e_3\},$$

(20.8) $$\{e_1, e_2, e_4\}.$$

In (20.6) suppress e_1; $\{e_2\}$ remains and it is a link. In (20.7) and (20.8) suppress e_1; there remain $\{e_2, e_3\}$ and $\{e_2, e_4\}$ which are links. Furthermore, e_2 is not useless, but e_3 and e_4 are. There is thus only one component that is not useless, and this network is equivalent to one that contains only e_2. This result may be seen more easily by considering cuts.

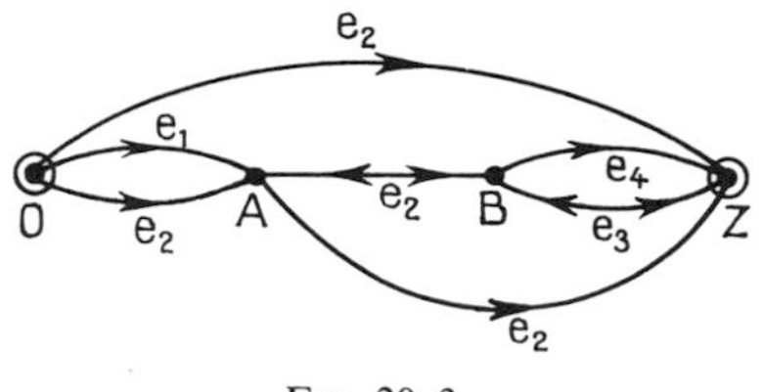

FIG. 20.3.

Equivalence between a Structure Function $\varphi(x)$ and a Reliability Network $\mathfrak{R}$. A reliability network $\mathfrak{R}$ and a structure function $\varphi(x)$ are said to be "equivalent," and we write

(20.9) $$\mathfrak{R} \simeq \varphi,$$

if, after suppressing any possible useless components, every link of $\mathfrak{R}$ is a link of $\varphi(x)$ and conversely (respectively, each cut of $\mathfrak{R}$ is a cut of $\varphi(x)$ and conversely).

Two equivalent structure functions (or two reliability networks or one reliability network and one structure function) correspond to the same system, taken to within useless components; they may be indifferently replaced with one another.

Remarks.

(1) The complementarity existing between links and cuts (the complementary subset of a link is not a cut; the complementary subset of a cut is not a link) evidently entails that, in the definition of equivalence, we may replace the word *link* by the word *cut* and conversely.

(2) The equivalence relation between reliability networks $\mathfrak{R}$ (respectively, between structure functions $\varphi(x)$) is evidently reflexive, transitive, and symmetric; it is indeed an equivalence relation in the sense of set theory.

Examples. The four reliability networks in Figs. 20.4–20.7 are equivalent. For the first three, the component e_2 is useless. The structure functions equivalent to these four networks are

$$\begin{aligned}\varphi_1(x) &= [1 - (1 - x_1)(1 - x_2)]\, x_1^2\,,\\ \varphi_2(x) &= [1 - (1 - x_1)(1 - x_2)]\, x_1\,,\\ \varphi_3(x) &= 1 - (1 - x_1)(1 - x_1 x_2)\,,\\ \varphi_4(x) &= x_1\,.\end{aligned}$$

These structure functions are equivalent to one another, as we have seen (see Eqs. (20.1)–(20.4)).

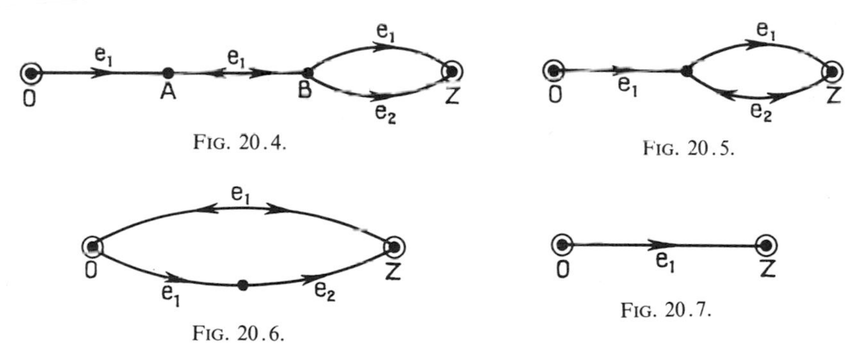

FIG. 20.4. FIG. 20.5. FIG. 20.6. FIG. 20.7.

21 Monotone (or Coherent) Structures

A structure function $\varphi(x)$ is monotone[10] if it possesses the following property:

$$(21.1)\qquad (x^{(2)}) \succcurlyeq (x^{(1)}) \;\Rightarrow\; \varphi(x^{(2)}) \geqslant \varphi(x^{(1)})\,.$$

[10] One also uses the term *coherent* to indicate that increasing monotone structures correspond to systems the design of which is normal, i.e., coherent.

Examples.

(1) Let

$$\varphi(x_1, x_2) = 1 - x_1 x_2 . \tag{21.2}$$

This structure function is not monotone. In fact, let

$$(x^{(1)}) = (x_1^{(1)}, x_2^{(1)}) = (0, 0), \tag{21.3}$$

$$(x^{(2)}) = (x_1^{(2)}, x_2^{(2)}) = (1, 1) . \tag{21.4}$$

We have

$$(1, 1) \succ (0, 0) . \tag{21.5}$$

On the other hand

$$\varphi(0, 0) = 1 , \qquad \varphi(1, 1) = 0 , \tag{21.6}$$

and

$$\varphi(1, 1) < \varphi(0, 0) ; \tag{21.7}$$

therefore (21.2) is not monotone.

(2) Let

$$\varphi(x_1, x_2) = 1 - (1 - x_1)(1 - x_2) . \tag{21.8}$$

This structure function is monotone. In fact

$$\varphi(0, 0) = 0 , \quad \varphi(0, 1) = 1 , \quad \varphi(1, 0) = 1 , \quad \varphi(1, 1) = 1 . \tag{21.9}$$

We therefore obtain

$$(1, 1) \succ (0, 1) \succ (0, 0) \quad \Rightarrow \quad \varphi(1, 1) = \varphi(0, 1) > \varphi(0, 0) , \tag{21.10}$$

$$(1, 1) \succ (1, 0) \succ (0, 0) \quad \Rightarrow \quad \varphi(1, 1) = \varphi(1, 0) > \varphi(0, 0) , \tag{21.11}$$

which satisfies (21.1).

(3) Let

$$\varphi(x_1, x_2) = x_1 x_2 . \tag{21.12}$$

This function is monotone, as one may easily check.

Theorem 21.I. *A structure function is monotone if and only if any set of components including a link is also a link.*

This theorem may be deduced immediately from the correspondence between r-tuples and subsets of components. If $\mathbf{a}_1$ and $\mathbf{a}_2$ are two subsets of components and $x^{(1)}$ and $x^{(2)}$ are r-tuples defined in the following fashion:

$$e_i \in \mathbf{a}_1 \quad \Rightarrow \quad x_i^{(1)} = 1 , \quad e_i \in \bar{\mathbf{a}}_1 \quad \Rightarrow \quad x_i^{(1)} = 0 , \tag{21.13}$$

$$e_i \in \mathbf{a}_2 \quad \Rightarrow \quad x_i^{(2)} = 1 , \quad e_i \in \bar{\mathbf{a}}_2 \quad \Rightarrow \quad x_i^{(2)} = 0 , \tag{21.14}$$

then

$$\mathbf{a}^{(2)} \supset \mathbf{a}^{(1)} \Leftrightarrow (x^{(2)}) \succcurlyeq (x^{(1)}) . \tag{21.15}$$

In fact, if all the components of $\mathbf{a}_1$ belong to $\mathbf{a}_2$, all the variables equal to 1 in $(x^{(1)})$ are also equal to 1 in $(x^{(2)})$ and conversely.

Relation (21.15) shows that, if a structure is monotone, that is, satisfies (21.1), it also satisfies the relation

$$\mathbf{a}_2 \supset \mathbf{a}_1 \Rightarrow \varphi(x^{(2)}) \geqslant \varphi(x^{(1)}) \tag{21.16}$$

where $\mathbf{a}_2$ and $x^{(2)}$ on one hand, $\mathbf{a}_1$ and $x^{(1)}$ on the other, are linked by relations (21.14) and (21.13). In particular, if $\mathbf{a}_1$ is a link, one has $\varphi(x^{(1)}) = 1$, therefore $\varphi(x^{(2)}) = 1$, and $\mathbf{a}_2$ is also a link. Conversely, if "$\mathbf{a}_1$ is a link" implies that "$\mathbf{a}_2$ is a link," then relation (21.1) is satisfied whenever $\varphi(x^{(1)}) = 1$. If, however, $\varphi(x^{(1)}) = 0$, it is satisfied automatically; it is thus true in all cases.

We thus recover for monotone structure functions the fundamental property of reliability networks announced in Theorem 19.I. It then follows that *monotone structure functions likewise satisfy Theorems* 19.II–19.IV. Furthermore, this result leads to the idea that there is an isomorphism (taken to within equivalence) between the set of reliability networks and the set of monotone structure functions. We shall see in fact in the following paragraphs that one may always construct a network equivalent to a monotone structure function and conversely. For the moment we may state the following result.

Theorem 21.II. *A structure function equivalent to a reliability network is monotone.*

Theorem 21.III. *A component of a reliability network $\mathfrak{R}$ or of a monotone structure function $\varphi(x)$ is useless if and only if it does not belong to any minimal link (respectively, to any minimal cut).*

If e_i is a useless component belonging to a link $\mathbf{a}$, the subset $\mathbf{a}' = \mathbf{a} - \{ e_i \}$ is again a link; e_i therefore does not belong to a minimal link. Conversely, if a component e_i belongs to no minimal link, let $\mathbf{a}$ be a link to which it belongs; this link contains a minimal link (Theorem 19.III), say $\mathbf{a}''$, which does not contain e_i. Any subset including $\mathbf{a}''$ is a link (Theorem 19.I or 21.I); in particular the subset $\mathbf{a}' = \mathbf{a} - \{ e_i \}$. This proves that any link containing e_i remains a link after suppression of e_i.

Theorem 21.IV. *Let $\varphi_1(x)$ and $\varphi_2(x)$ be monotone structure functions and $\mathfrak{R}_1$ and $\mathfrak{R}_2$ reliability networks. There is an equivalence among these four entities if they have the same set of minimal links (respectively, the same set of minimal cuts).*

In fact, it follows from Theorem 21.III that two "entities" having the same set of minimal links have the same set of nonuseless components. On the other hand, Theorems 19.III and 19.I or 21.I show that the set of links of a monotone structure function defined on a set of components **e** is the set of subsets of **e** including a minimal link; the identity of sets of minimal links then entails the identity of the set of links, after suppression of useless components.

Definition *The order of a reliability network* $\mathfrak{R}$ *or of a structure function* $\varphi(x)$ *is the number of its nonuseless components. A degenerate structure is of order* 0.

Theorem 21.V [28]. *Given a structure function* $\varphi(x)$ *that possesses n components, let*

(21.17) A_k *be the number of links having k components;*

(21.18) B_k *be the number of cuts having k components;*

(21.19) A'_k *be the number of minimal links having k components;*

(21.20) $\mathscr{B}^i_k$ *be the number defined in the following fashion: Let* $\mathbf{b}_1^{n-k}, \mathbf{b}_2^{n-k}, \ldots, \mathbf{b}_{B_{n-k}}^{n-k}$ *be the cuts having* $n - k$ *components. For arbitrary i* $(1 \leqslant i \leqslant B_{n-k})$, *the complementary subset* $\overline{\mathbf{b}_i^{n-k}}$ *is then not a link; it has k components. Consider the* $n - k$ *sets obtained by adding to* $\overline{\mathbf{b}_i^{n-k}}$ *a component that does not already appear there;* $\mathscr{B}^i_k$ *is the number of these sets that are nonminimal links.*

A necessary and sufficient condition for the structure $\varphi(x)$ *to be monotone is that one have, for all values of k such that* $0 \leqslant k \leqslant n$:

$$(n - k)\, A_k + \sum_{i=1}^{B_{n-k}} \mathscr{B}^i_k = (k + 1)\,(A_{k+1} - A'_{k+1}) \tag{21.21}$$

or the equivalent equality

$$kB_k = (n - k + 1)\,(A'_{n-k+1} + B_{k-1}) + \sum_{i=1}^{B_k} \mathscr{B}^i_{n-k}\,. \tag{21.22}$$

Equation (21.21) expresses the fact that by adding an arbitrary component to the links having k components and to complementaries of certain cuts having $n - k$ components, one obtains all the nonminimal links having $k + 1$ components.

Equation (21.22) states that by suppressing an arbitrary component of the cuts having k components, one obtains all the cuts having $k - 1$ components, all minimal links having $n - k + 1$ components, and the nonminimal links having $n - k + 1$ components obtained through (21.20). For proof of this theorem and of the two following, we refer the reader to the work of Hansel [28].

Example. Consider a structure function of order 4:

$$\varphi(x) = 1 - (1 - x_1 x_2)(1 - x_3)(1 - x_1 x_4) . \tag{21.23}$$

x_1	x_2	x_3	x_4	φ
0	0	0	0	0
0	0	0	1	0
0	0	1	0	1
0	0	1	1	1
0	1	0	0	0
0	1	0	1	0
0	1	1	0	1
0	1	1	1	1
1	0	0	0	0
1	0	0	1	1
1	0	1	0	1
1	0	1	1	1
1	1	0	0	1
1	1	0	1	1
1	1	1	0	1
1	1	1	1	1

FIG. 21.1.

By constructing the table of values of this function (Fig. 21.1), one may verify that it is monotone. Moreover, all functions of the form $1 - \prod(1 - x_i x_j \cdots x_l)$ are monotone since their derivatives, of the form

$$\sum x_i x_j \cdots x_k \prod(1 - x_m x_n \cdots x_q)$$

are always positive. This table also permits the recording of the links and cuts, and the computation of their numbers of components (see Fig. 21.2).

We calculate for this example the numbers $\mathcal{B}_k^i$ defined by (21.20) with the aid of Figs. 21.1 and 21.2, which represents the lattice of states of the set of components (cf. Section 15, Fig. 15.2):

(a) $k = 0$. The only cut having $n - k = 4$ components is $\mathbf{b}_1^4 = \{e_1, e_2, e_3, e_4\}$. The complementary subset $\overline{\mathbf{b}_1^4}$ is the empty set; by adding to it one component, one obtains the sets $\{e_1\}$, $\{e_2\}$, $\{e_3\}$, and $\{e_4\}$, among which $\{e_3\}$ is the only link, but it is a minimal link. Thus $\mathcal{B}_0^1 = 0$.

(b) $k = 1$. $B_{n-k} = B_3 = 3$; $\mathbf{b}_1^3 = \{e_1, e_2, e_3\}$; $\overline{\mathbf{b}_1^3} = \{e_4\}$. Among the sets $\{e_1, e_4\}$, $\{e_2, e_4\}$, and $\{e_3, e_4\}$, only $\{e_3, e_4\}$ is a nonminimal link. Thus $\mathcal{B}_1^1 = 1$. Then $\mathbf{b}_2^3 = \{e_1, e_3, e_4\}$, $\overline{\mathbf{b}_2^3} = \{e_2\}$; among the vertices of the lattice joined to vertex (0100), only (0110) corresponds to a nonminimal link $\{e_2, e_3\}$. Thus $\mathcal{B}_1^2 = 1$. Finally, $\mathbf{b}_3^3 = \{e_2, e_3, e_4\}$. Thus $\mathcal{B}_1^3 = 1$.

(c) $k = 2$. $B_2 = 1$; $\mathbf{b}_1^2 = \{e_1, e_3\}$; $\overline{\mathbf{b}_1^2} = \{e_2, e_4\}$. Thus $\mathcal{B}_2^1 = 2$.

(d) $k = 1$ and $k = 0$. $B_1 = B_0 = 0$.

Links		Cuts	
Number of components	Number of links	Number of components	Number of cuts
4	1	0	0
3	4	1	0
2	5	2	1
1	1	3	3
0	0	4	1

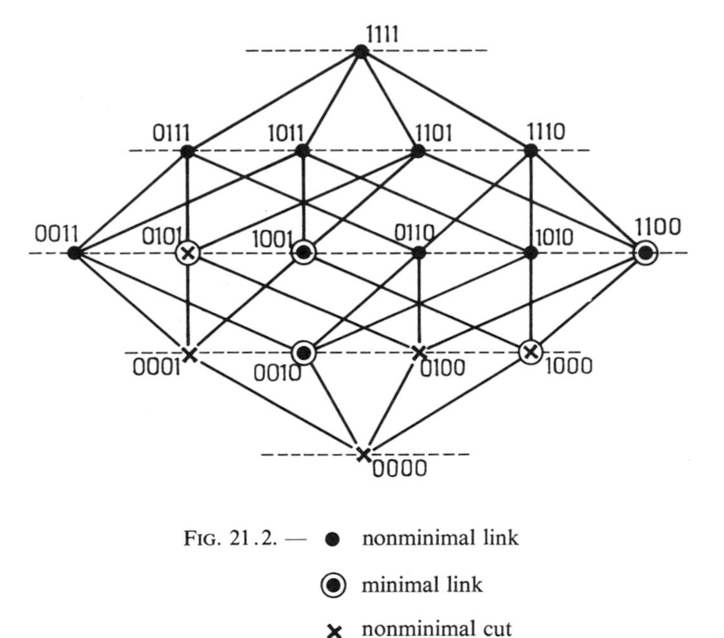

FIG. 21.2. — ● nonminimal link

◉ minimal link

× nonminimal cut

⊗ minimal cut

From these results, it is easy to verify (21.21) and (21.22). For example, for (21.21) and $k = 2$, we have $A_2 = 5$, $\sum \mathcal{B}_2^i = 2$, $A_3 = 4$, and $A'_3 = 0$, from which

$$(4 - 2) \times 5 + 2 = 3 \times (4 - 0).$$

From Theorem 21.V we immediately deduce the following property.

Theorem 21.VI. *A monotone structure satisfies the following inequalities:*

$$(21.24) \qquad (k + 1)\, A_{k+1} \geqslant (n - k)\, A_k\,, \qquad k = 0, 1, 2, \ldots, n - 1\,,$$

and the equivalent inequalities

$$\frac{A_{k+1}}{\binom{n}{k+1}} \geqslant \frac{A_k}{\binom{n}{k}}, \tag{21.25}$$

$$kB_k \geqslant (n - k + 1)\, B_{k-1}\,. \tag{21.26}$$

We note that these inequalities are not sufficient to assure the monotonicity of a structure $\varphi(x)$.

Theorem 21.VII [30]. *The number $\psi(n)$ of monotone structures of order less than or equal to n satisfies the following inequalities:*

$$2^{\binom{n}{\langle n/2 \rangle}} \leqslant \psi(n) \leqslant 3^{\binom{n}{\langle n/2 \rangle}} \tag{21.27}$$

where $\langle n/2 \rangle$ is the largest integer less than or equal to $n/2$.

From (21.27) we deduce

$$\begin{aligned} 4 &\leqslant \psi(2) \leqslant 9\,, \\ 8 &\leqslant \psi(3) \leqslant 27\,, \\ 64 &\leqslant \psi(4) \leqslant 729\,, \\ 1\,024 &\leqslant \psi(5) \leqslant 59\,049\,, \\ 1\,048\,576 &\leqslant \psi(6) \leqslant 3.481 \times 10^9\,. \end{aligned}$$

The combinatorial variety of structures thus becomes very large when the number of components exceeds 5. We shall examine in Section 26 how to enumerate these structures.

The exact values of $\psi(n)$ have been calculated by Gilbert [24] for $n < 7$. He obtains

$$\psi(1) = 3\,, \quad \psi(2) = 6\,, \quad \psi(3) = 20\,, \quad \psi(4) = 168\,,$$
$$\psi(5) = 7\,580\,, \quad \psi(6) = 7\,828\,354\,.$$

Equivalence between Monotone Structures and the Free Distributive Lattice on n Generators[11]**.** One may easily show the isomorphism existing between the set of monotone structure functions having n components provided with the series composition[12] and parallel composition operations and

[11] The reader who wishes to study lattice theory in more depth should see, for example, A. Kaufmann and M. Précigout, *Cours de Mathématiques Nouvelles pour le Recyclage des Ingénieurs,* Dunod, Paris, 1966; R. Faure, A. Kaufmann, and M. Denis-Papin, *Cours de Calcul Booléien Appliqué*, Albin-Michel, Paris, 1963.

[12] A complete study of the composition of structure function and reliability networks is given in Section 26.

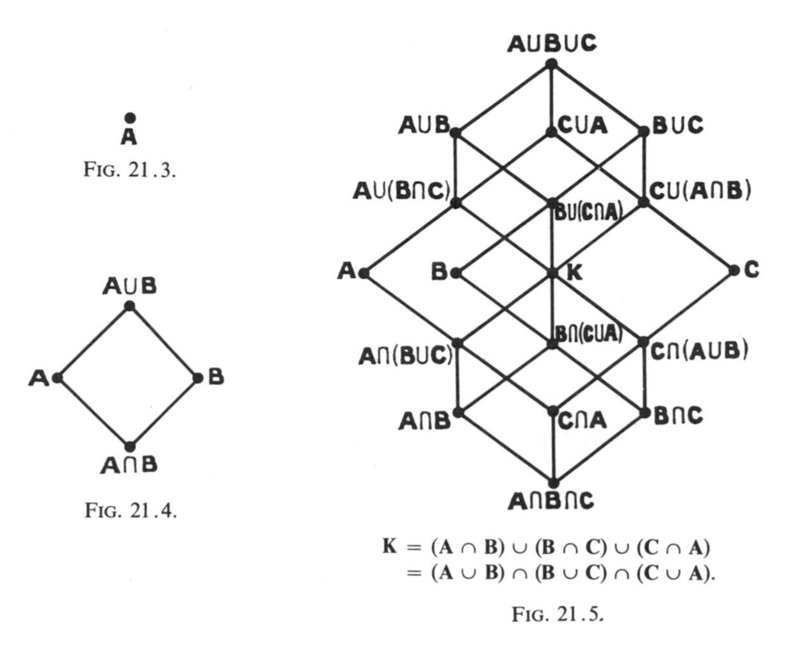

FIG. 21.3.

FIG. 21.4.

$$\mathbf{K} = (\mathbf{A} \cap \mathbf{B}) \cup (\mathbf{B} \cap \mathbf{C}) \cup (\mathbf{C} \cap \mathbf{A})$$
$$= (\mathbf{A} \cup \mathbf{B}) \cap (\mathbf{B} \cup \mathbf{C}) \cap (\mathbf{C} \cup \mathbf{A}).$$

FIG. 21.5.

the free distributive lattice on n generators for the operations intersection and union (the ordered set of Boolean functions that may be written without using the operation of complementation). Figures 21.3–21.5 represent the free distributive lattices on n generators for $n = 1, 2, 3$; these lattices are represented by their Hasse diagrams. The isomorphism indicated above does not include the degenerate functions $\varphi(x) \equiv 1$ and $\varphi(x) \equiv 0$, otherwise it would be necessary to enrich the free distributive lattice on n generators with the element **c** (the reference) and $\varnothing$ (the empty set).

One may arbitrarily associate

series composition ⇔ intersection
parallel composition ⇔ union

or

series composition ⇔ union
parallel composition ⇔ intersection.

We give in Figs. 21.6–21.8 the reliability networks corresponding to monotone functions for $n = 1$, 2, and 3.[13] The networks corresponding to the degenerate functions $\varphi(x) \equiv 1$ and $\varphi(x) \equiv 0$ have not been represented. The positions of the lattice elements correspond to those of Figs. 21.3–21.5. On the other hand, the networks have been represented by nonoriented graphs, the components being indicated with the aid of their respective state variables.

[13] We have used the first of these two conventions above.

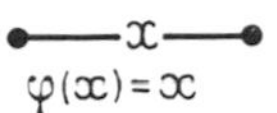

FIG. 21.6. Lattice of monotone structure function with one component (trivial)

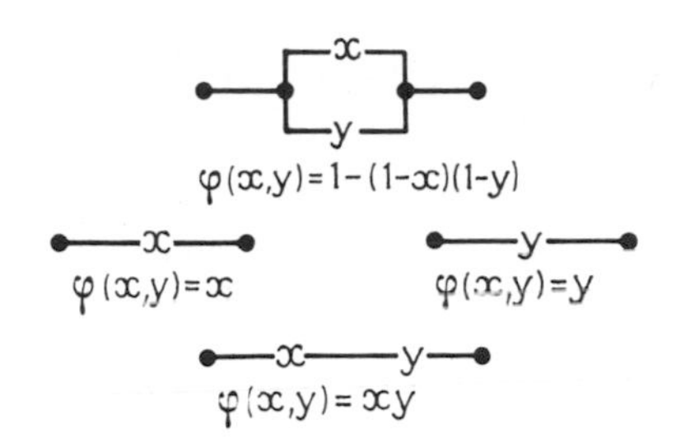

FIG. 21.7. Lattice of monotone structure functions with two components.

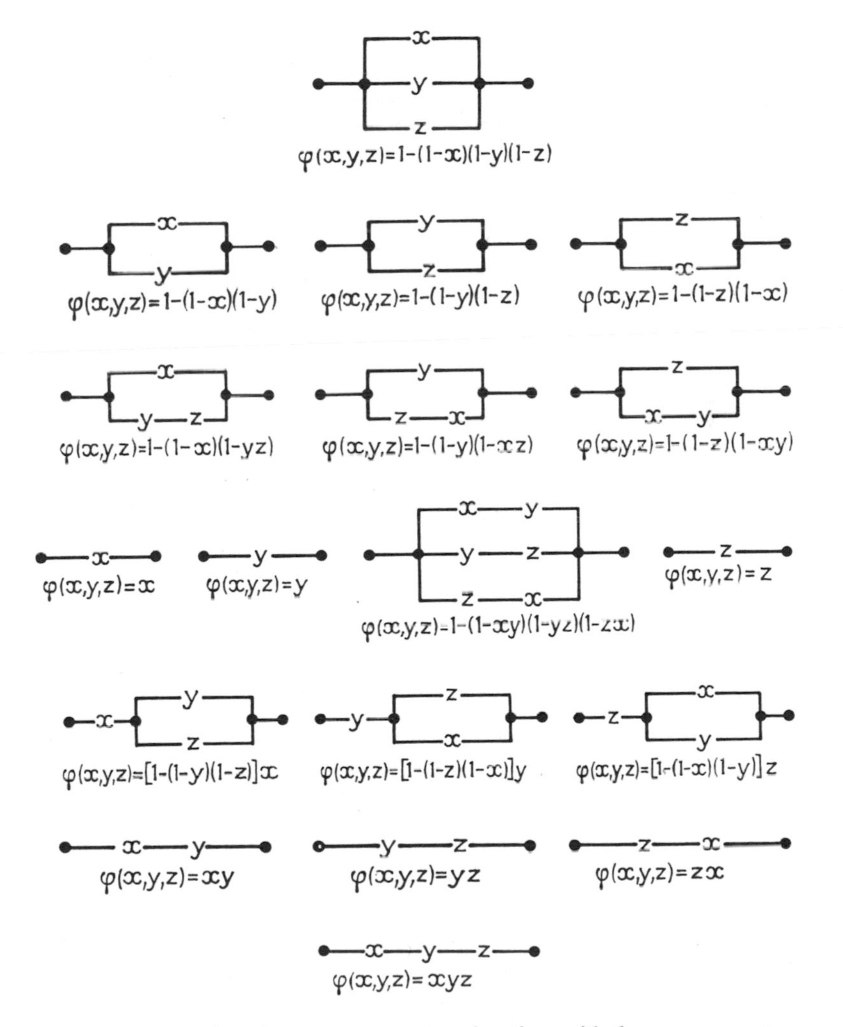

FIG. 21.8. Lattice of monotone structure functions with three components.

22 Construction and Simplification of Structure Functions and of Reliability Networks

Construction of a Monotone Structure Function Equivalent to a Network

A. Method of Links. Suppose that we have determined all the minimal links $\mathbf{a}_1, \mathbf{a}_2, \ldots, \mathbf{a}_k$ of a network $\mathcal{R}$, and let x_i be the state variable of the component e_i; the structure function

$$\varphi(x) = 1 - \prod_{j=1}^{k} \left(1 - \prod_{\substack{i \\ e_i \in \mathbf{a}_j}} x_i\right) \tag{22.1}$$

is equivalent to the network $\mathcal{R}$.

In fact, any minimal link of $\mathcal{R}$ is a link of $\varphi(x)$: if $x_i = 1$ for any $e_i \in \mathbf{a}_j$, then $\prod_{i:e_i \in \mathbf{a}_j} x_i = 1$ and therefore $\prod_{j=1}^{k}(1 - \prod_{i:e_i \in \mathbf{a}_j} x_i) = 0$, and then $\varphi(x) = 1$.

Similarly, a minimal link of $\varphi(x)$ necessarily cancels at least one factor $1 - \prod_{i:e_i \in \mathbf{a}_j} x_i$, that is, it includes at least one minimal link $\mathbf{a}_j$ of $\mathcal{R}$; this is therefore also a link of $\mathcal{R}$. It follows that φ and $\mathcal{R}$ have the same set of minimal links and are equivalent according to Theorem 21.IV.

Formula (22.1) remains valid if $\{ \mathbf{a}_1, \mathbf{a}_2, \ldots, \mathbf{a}_k \}$ designates a set of links including all minimal links of $\mathcal{R}$, and perhaps certain nonminimal links. This remark will allow us to use a convenient method called "the method of routes." In Section 23 we shall give a general exposition of this method.

Example (Fig. 22.1). The network of this figure possesses three paths from O to Z:

$$\{ (O, C, B, D, Z), (O, C, B, A, Z), (O, A, Z) \} . \tag{22.2}$$

To these paths correspond the following links, which are minimal:

$$\{ \{ e_1, e_2, e_3 \}, \{ e_1, e_2, e_4 \}, \{ e_3, e_4 \} \} . \tag{22.3}$$

We therefore have

$$\varphi(x_1, x_2, x_3, x_4) = 1 - (1 - x_1 x_2 x_3)(1 - x_1 x_2 x_4)(1 - x_3 x_4) . \tag{22.4}$$

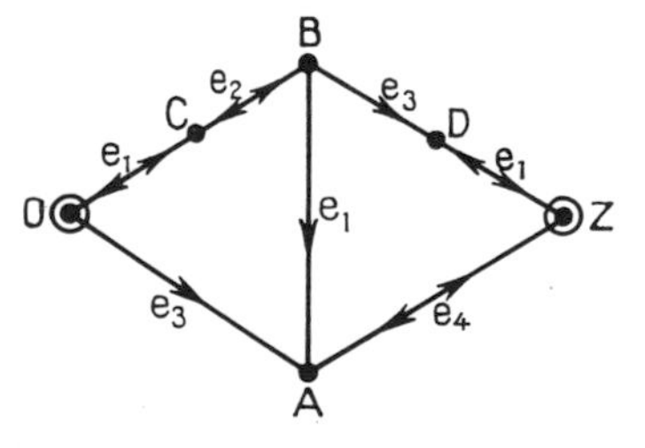

FIG. 22.1.

B. Method of Cuts. Suppose that we have determined a set of cuts $\{ \mathbf{b}_1, \mathbf{b}_2, \ldots, \mathbf{b}_k \}$ including all minimal cuts and perhaps some nonminimal cuts of a reliability network $\mathfrak{R}$. Then the structure function

$$\varphi(x) = \prod_{j=1}^{k} \left[1 - \prod_{\substack{i \\ e_i \in \mathbf{b}_j}} (1 - x_i) \right] \tag{22.5}$$

is equivalent to $\mathfrak{R}$. This is proved using reasoning similar to that used for (22.1).

Example (Fig. 22.1). By examining this network we easily see that there are five minimal cuts:

$$\{ \{ e_1, e_3 \}, \{ e_1, e_4 \}, \{ e_2, e_3 \}, \{ e_2, e_4 \}, \{ e_3, e_4 \} \} . \tag{22.6}$$

We therefore have

$$\begin{aligned} \varphi(x_1, x_2, x_3, x_4, x_5) &= [1 - (1 - x_1)(1 - x_3)] \\ &\cdot [1 - (1 - x_1)(1 - x_4)] \cdot [1 - (1 - x_2)(1 - x_3)] \\ &\cdot [1 - (1 - x_2)(1 - x_4)] \cdot [1 - (1 - x_3)(1 - x_4)] . \end{aligned} \tag{22.7}$$

By carrying out the products we may verify that the two functions (22.4) and (22.7) are equivalent to

$$x_1 x_2 x_3 + x_1 x_2 x_4 + x_3 x_4 - 2x_1 x_2 x_3 x_4 .$$

Construction of a Network Equivalent to a Monotone Structure Function. The methods described above are still valid in principle. Knowing a set of links $\{ \mathbf{a}_1, \mathbf{a}_2, \ldots, \mathbf{a}_k \}$ including all the minimal links of a monotone structure function, we obtain an equivalent network by placing in parallel k subnetworks, each formed of the components of a link placed in series (Fig. 22.2).[14]

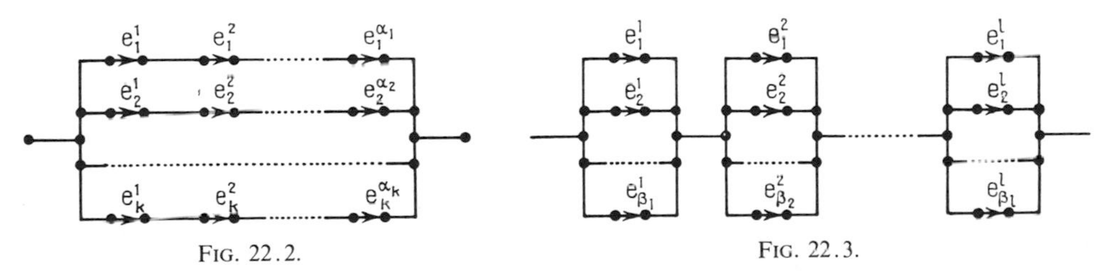

FIG. 22.2. FIG. 22.3.

Similarly, knowing a set of cuts $\{ \mathbf{b}_1, \mathbf{b}_2, \ldots, \mathbf{b}_l \}$ including all minimal cuts of a monotone structure function, we obtain an equivalent network by placing these cuts in series (Fig. 22.3)[14] each having its components in parallel.

[14] In Figs. 22.2 and 22.3 we have placed arrowheads only in the direction from entry to exit; given the nature of these networks, one would obtain equivalent networks by freely adding arrowheads in the opposite sense.

Example. Consider the structure function

(22.8) $$\varphi(x_1, x_2, x_3) = x_1\, x_3 + x_2\, x_3 - x_1\, x_2\, x_3\,.$$

The table of values of $\varphi(x_1, x_2, x_3)$ is shown in Fig. 22.4.

	x_1	x_2	x_3	$x_1\,x_3$	$x_2\,x_3$	$x_1\,x_2\,x_3$	$x_1\,x_3 + x_2\,x_3 - x_1\,x_2\,x_3$
(0)	0	0	0	0	0	0	0
(1)	0	0	1	0	0	0	0
(2)	0	1	0	0	0	0	0
(3)	0	1	1	0	1	0	1
(4)	1	0	0	0	0	0	0
(5)	1	0	1	1	0	0	1
(6)	1	1	0	0	0	0	0
(7)	1	1	1	1	1	1	1

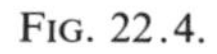

FIG. 22.4.

We should verify first that this function is indeed monotone, which is easy by examining the pairs of rows i and j for the dominance order relation. This function is indeed monotone. Now, by examining rows 3, 5, and 7, we see that there are three links:

$$\{\, e_2, e_3 \,\},\ \{\, e_1, e_3 \,\},\ \{\, e_1, e_2, e_3 \,\}\,.$$

Thus, an equivalent network is given in Fig. 22.5. Similarly, by examining rows 0, 1, 2, 4, and 6, we see that there are five cuts

$$\{\, e_1, e_2 \,\},\ \{\, e_1, e_3 \,\},\ \{\, e_2, e_3 \,\},\ \{\, e_3 \,\},\ \{\, e_1, e_2, e_3 \,\}\,.$$

Thus, an equivalent network is given in Figure 22.6.

In Figs. 22.5 and 22.6 we have purposefully not used the notions of minimal links and cuts: we shall see a little later how to simplify such networks.

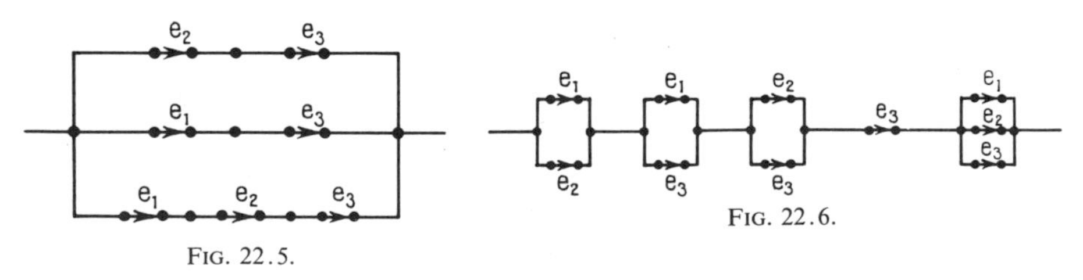

FIG. 22.5.

FIG. 22.6.

Remark. The construction procedures developed above show that it is always possible to determine a monotone structure function equivalent to a network, or a network equivalent to a monotone structure function, a result that we have announced in the comments on Theorem 21.I.

Simplification of a Monotone Structure Function. Simple Form. Let $\varphi(x)$ be a monotone structure function for which we suppose known a set of links $\{ \mathbf{a}_1, \mathbf{a}_2, \dots, \mathbf{a}_k \}$ including all minimal links. Formula (22.1) then gives

$$\varphi(x) = 1 - \prod_{j=1}^{k} \left(1 - \prod_{e_i \in \mathbf{a}_j} x_i\right). \tag{22.9}$$

Here $\varphi(x)$ is a polynomial with respect to the Boolean variables x_i; since x_i may take only the values 0 or 1, one therefore has

$$(x_i)^r = x_i\,, \qquad r = 1, 2, 3, \dots . \tag{22.10}$$

One concludes that $\varphi(x)$ may always be written in the form of a polynomial of first degree with respect to each variable x_i. We shall call "simple form," denoted $\varphi_s(x)$, the function that is equivalent to $\varphi(x)$ and that is a sum of distinct monomials, each of first degree with respect to each x_i. It is easy to prove that any simple form is unique.

Example. Let

$$\begin{aligned}
\varphi(x_1, x_2, x_3) &= 1 - (1 - x_1 x_2)(1 - x_1 x_3)(1 - x_2 x_3) \\
&= 1 - [1 - x_1 x_2 - x_1 x_3 + x_1^2 x_2 x_3](1 - x_2 x_3) \\
&= 1 - [1 - x_1 x_2 - x_1 x_3 + x_1 x_2 x_3][1 - x_2 x_3] \\
&= x_1 x_2 + x_1 x_3 - x_1 x_2 x_3 + x_2 x_3 - x_1 x_2^2 x_3 \\
&\quad - x_1 x_2 x_3^2 + x_1 x_2^2 x_3^2 \\
&= x_1 x_2 + x_1 x_3 - x_1 x_2 x_3 + x_2 x_3 - x_1 x_2 x_3 \\
&\quad - x_1 x_2 x_3 + x_1 x_2 x_3 \\
&= x_1 x_2 + x_1 x_3 + x_2 x_3 - 2 x_1 x_2 x_3 .
\end{aligned} \tag{22.11}$$

We therefore have

$$\varphi_s(x) = x_1 x_2 + x_1 x_3 + x_2 x_3 - 2 x_1 x_2 x_3 . \tag{22.12}$$

Construction and Simplification of Nonmonotone Structure Functions. Let $\varphi(x)$ be a nonmonotone structure function and $(x^{(1)}), (x^{(2)}), \dots, (x^{(k)})$ the states of the set of components such that $\varphi(x^{(i)}) = 1$, $i = 1, 2, \dots, k$, that is, the states to which correspond links (cf. Theorem 17.I). Define:

$\mathbf{m}_i$ as the set of components in a good state (state variables equal to 1 in $(x^{(i)})$), which constitute the link associated with $x^{(i)}$;

$\mathbf{m}_i$ as the set of failed components (state variables equal to 0 in $(x^{(i)})$);
$\overline{\mathbf{m}}_i$ as the complementary set of $\mathbf{m}_i$, that is, $\overline{\mathbf{m}}_i = \mathbf{e} - \mathbf{m}_i$.

One may then easily show that[15]

$$\varphi(x) = 1 - \prod_{i=1}^{k} \left[1 - \prod_{e_j \in \mathbf{m}_i} x_j \cdot \prod_{e_j \in \overline{\mathbf{m}}_i} (1 - x_j)\right]. \tag{22.13}$$

Similarly, if $(x^{(k+1)}), (x^{(k+2)}), \ldots, (x^{(l)})$ are states such that $\varphi(x^{(i)}) = 0$, $i = k+1, k+2, \ldots, l$, where $l = 2^r$ is the total number of states, define:

$\mathbf{n}_i$ as the set of failed components of $(x^{(i)})$ (state variables equal to 0 of $(x^{(i)})$), which constitute the cut associated with $x^{(i)}$; and
$\overline{\mathbf{n}}_i$ as the complementary set $\overline{\mathbf{n}}_i = \mathbf{e} - \mathbf{n}_i$.

One may similarly show, with the same convention of notation as for (22.13), that

$$\varphi(x) = \prod_{i=k+1}^{l} \left[1 - \prod_{e_j \in \mathbf{n}_i} (1 - x_j) . \prod_{e_j \in \overline{\mathbf{n}}_i} x_j\right]. \tag{22.14}$$

Expressions (22.13) and (22.14) are similar, respectively, to (22.1) and (22.5), but two important differences are to be noted:

(a) In (22.13), the product must be taken extended to all r-tuples for which $\varphi = 1$, that is, to all links, whereas in (22.1) it was sufficient to consider minimal links; similarly, in (22.14) it is necessary to consider all r-tuples for which $\varphi = 0$.

(b) Each variable appears once, either in the form x_j or in the form $1 - x_j$ in each of the factors of the products of (22.13) and (22.14).

Expressions (22.1) and (22.5) are simpler since they use the fundamental property (cf. Theorem 21.I) of monotone structures, that is, that any set of components including a link (respectively, a cut) is likewise a link (respectively, a cut).

There exist, however, in the general case of not necessarily monotone structures, some more convenient formulas than (22.13) and (22.14); we shall see in Section 26 that these are related to the notion of linear composition of structures, but one may also obtain them from the relation

$$\sum_{i=1}^{l} \left[\prod_{e_j \in \mathbf{m}_i} x_j \cdot \prod_{e_j \in \overline{\mathbf{m}}_i} (1 - x_j)\right] = 1 \tag{22.15}$$

where $\mathbf{m}_i$ designates, as in (22.13), the set of state variables x_j equal to 1 for a

[15] If one of the subsets $\mathbf{m}_i$ or $\overline{\mathbf{m}}_i$ is empty, the corresponding product Πx_j or $\Pi(1 - x_j)$ will be taken equal to 1.

given r-tuple i. (Note that in the notations of (22.14) one has $\mathbf{n}_i = \overline{\mathbf{m}}_i$.) This relation is very easy to prove: if one considers an r-tuple

$$\xi^{(i)} = (\xi_1^{(i)}, \xi_2^{(i)}, \ldots, \xi_r^{(i)}), \tag{22.16}$$

with

$$\begin{aligned} \xi_j^{(i)} &= 1 \quad \text{if} \quad e_j \in \mathbf{m}_i \\ &= 0 \quad \text{if} \quad e_j \in \overline{\mathbf{m}}_i, \end{aligned} \tag{22.17}$$

the ith term of the first member of (22.15) will be equal to 1 when one puts $x_j = \xi_j^{(i)}$, and all the other terms will be zero.

It follows from (22.15) that one may write

$$\varphi(x) = \sum_{i=1}^{k} \left[\prod_{e_j \in \mathbf{m}_i} x_j \cdot \prod_{e_j \in \overline{\mathbf{m}}_i} (1 - x_j) \right] \tag{22.18}$$

and also

$$\begin{aligned} \varphi(x) &= 1 - \sum_{i=k+1}^{l} \left[\prod_{e_j \in \mathbf{m}_i} x_j \cdot \prod_{e_j \in \overline{\mathbf{m}}_i} (1 - x_j) \right] \\ &= 1 - \sum_{i=k+1}^{l} \left[\prod_{e_j \in \mathbf{n}_i} (1 - x_j) \cdot \prod_{e_j \in \overline{\mathbf{n}}_i} x_j \right]. \end{aligned} \tag{22.19}$$

These two relations indeed give $\varphi = 1$ for $(x) = \xi^{(i)}$, $i = 1, \ldots, k$, and $\varphi = 0$ for $(x) = \xi^{(i)}$, $i = k+1, \ldots, l$. These have the advantage over (22.13) and (22.14) of being of first degree with respect to each of the variables; passage to the simple form is therefore more rapid.

Example 1. Let

$$\varphi_s(x_1, x_2, x_3) = 1 + x_1 x_2 x_3 - x_1 . \tag{22.20}$$

We construct a table allowing us to obtain $\varphi(x_1, x_2, x_3)$ for the eight possible states (Fig. 22.7). We may verify that this function is not monotone by comparing, for example, row 5 with row 1.

We see that $\varphi(x_1, x_2, x_3) = 1$ for rows 0, 1, 2, 3, and 7, for which we have

$$\begin{aligned} &\mathbf{m}_0 = \varnothing, \quad \overline{\mathbf{m}}_0 = \{e_1, e_2, e_3\}, \quad \mathbf{m}_1 = \{e_3\}, \quad \overline{\mathbf{m}}_1 = \{e_1, e_2\}, \\ &\mathbf{m}_2 = \{e_2\}, \quad \overline{\mathbf{m}}_2 = \{e_1, e_3\}, \quad \mathbf{m}_3 = \{e_2, e_3\}, \quad \overline{\mathbf{m}}_3 = \{e_1\}, \\ &\mathbf{m}_7 = \{e_1, e_2, e_3\}, \quad \overline{\mathbf{m}}_7 = \varnothing . \end{aligned}$$

Formula (22.18) gives

(22.21)

$$\begin{aligned} \varphi(x_1, x_2, x_3) = {} & (1 - x_1)(1 - x_2)(1 - x_3) + (1 - x_1)(1 - x_2)\, x_3 \\ & + (1 - x_1)\, x_2 (1 - x_3) + (1 - x_1)\, x_2 x_3 + x_1 x_2 x_3 . \end{aligned}$$

	x_1	x_2	x_3	$x_1\,x_2\,x_3$	$1 - x_1$	φ
(0)	0	0	0	0	1	1
(1)	0	0	1	0	1	1
(2)	0	1	0	0	1	1
(3)	0	1	1	0	1	1
(4)	1	0	0	0	0	0
(5)	1	0	1	0	0	0
(6)	1	1	0	0	0	0
(7)	1	1	1	1	0	1

FIG. 22.7

Formula (22.19) gives

(22.22)

$$\varphi(x_1, x_2, x_3) = 1 - x_1(1 - x_2)\,(1 - x_3) - x_1(1 - x_2)\,x_3 - x_1\,x_2(1 - x_3)\,.$$

The reader should be able to verify easily that the two functions (22.21) and (22.22) are equivalent to (22.20).

By way of comparison, formula (22.14) would give a clearly more complicated expression:

(22.23)

$$\varphi(x_1, x_2, x_3) = [1 - (1 - x_2)\,(1 - x_3)\,x_1] \cdot [1 - (1 - x_1)\,(1 - x_3)\,x_2].[1 - (1 - x_1)\,(1 - x_2)\,x_3]\,.$$

Example 2. Let

(22.24) $$\varphi(x_1, x_2, x_3) = (1 - x_1\,x_2\,x_3)\,(1 - x_2\,x_3)\,x_1\,x_3\,.$$

	x_1	x_2	x_3	$x_2\,x_3$	$x_1\,x_2\,x_3$	$1 - x_1\,x_2\,x_3$	$1 - x_2\,x_3$	$x_1\,x_3$	φ
(0)	0	0	0	0	0	1	1	0	0
(1)	0	0	1	0	0	1	1	0	0
(2)	0	1	0	0	0	1	1	0	0
(3)	0	1	1	1	0	1	0	0	0
(4)	1	0	0	0	0	1	1	0	0
(5)	1	0	1	0	0	1	1	1	1
(6)	1	1	0	0	0	1	1	0	0
(7)	1	1	1	1	1	0	0	1	0

FIG. 22.8.

In Fig. 22.8 we show a table allowing one to obtain $\varphi(x_1, x_2, x_3)$ for the eight distinct states.

We may verify that this function is not monotone (see rows 7 and 5). We see that $\varphi(x_1, x_2, x_3) = 1$ only for row 5. We have

$$\mathbf{m}_5 = \{ e_1, e_3 \}, \qquad \overline{\mathbf{m}}_5 = \{ e_2 \}, \tag{22.25}$$

from which

$$\begin{aligned} \varphi(x_1, x_2, x_3) &= x_1 x_3 (1 - x_2) \\ &= x_1 x_3 - x_1 x_2 x_3 . \end{aligned} \tag{22.26}$$

We may check this by comparing Figs. 22.8 and 22.9.

x_1	x_2	x_3	$x_1 x_3$	$x_1 x_2 x_3$	φ
0	0	0	0	0	0
0	0	1	0	0	0
0	1	0	0	0	0
0	1	1	0	0	0
1	0	0	0	0	0
1	0	1	1	0	1
1	1	0	0	0	0
1	1	1	1	1	0

FIG. 22.9.

Remark. We do not know of a general method to obtain directly the simple form of a structure function, be it monotone or not, or a reliability network with minimum numbers of arcs or vertices. However, interesting results have been obtained in some particular cases, notably for "k of n" structures (see Section 35).

23 Finding Links and Cuts

Purpose of This Search. As we have seen in the preceding sections, the determination of a structure equivalent to a network is simple if one knows a set of links including all the minimal links, or a set of cuts including all the minimal cuts of the network being considered.

Conversely, knowledge of a set of links (or cuts) including all the minimal links (or cuts) of a monotone structure function permits one to determine easily an equivalent network. We shall now describe several simple algorithms allowing one to determine such sets including all minimal links or all minimal cuts.

Search for a Set of Links Including All Minimal Links of a Network. Recall that a reliability network $\mathfrak{R}$ has been defined in Section 19 as an r-fold graph $G = (\mathbf{S}, \mathbf{U})$ and a mapping Δ of the set $\mathbf{U}$ of arcs of the graph into the set $\mathbf{e}$ of components of the system. To a link $\mathbf{a}$ is associated a partial graph $G_p(\mathbf{a})$, in which occur only the arcs u_j such that $\Delta(u_j) \in \mathbf{a}$, and such that there exists in $G_p(\mathbf{a})$ a path from O to Z. Then, however, there exists in $G_p(\mathbf{a})$ an elementary path μ from O to Z (cf. Section 18). Conversely, to such an elementary path $\mu = (u_1, \dots, u_l)$ one may associate a link $\mathbf{a}' = \{ \Delta(u_1) \} \cup \cdots \cup \{ \Delta(u_l) \}$[16]; it is clear that $\mathbf{a}' \subset \mathbf{a}$. It follows that, if one considers the set $\mathfrak{C}$ of elementary paths of the graph connecting O to Z, one may associate with it a set $\mathfrak{L}$ of links such that each link of the network includes a link of the set $\mathfrak{L}$. A minimal link including no links other than itself then necessarily belongs to $\mathfrak{L}$. This result, which is at the foundation of the search methods for links of a network, is expressed in the following theorem.

Theorem 23.1. *Let $\mathfrak{C}$ be the set of elementary paths with initial end O and terminal end Z of the graph G of a reliability network. To each path $\mu = (u_1, \dots, u_l) \in \mathfrak{C}$ associate the link $\mathbf{a}(\mu)$ formed of the components corresponding to the arcs of the path*

$$\mathbf{a}(\mu) = \{ \Delta(u_1) \} \cup \cdots \cup \{ \Delta(u_j) \} \cup \cdots \cup \{ \Delta(u_l) \} . \tag{23.1}$$

Let $\mathfrak{L}$ be the set of links thus obtained from $\mathfrak{C}$[17]:

$$\mathfrak{L} = \bigcup_{\mu \in \mathfrak{C}} \{ \mathbf{a}(\mu) \} . \tag{23.2}$$

Then the set $\mathfrak{L}$ includes all the minimal links of the network.

Example. The elementary paths of the network in Fig. 23.1 are

$$(OB, BC, CZ), \quad (OB, BC, CA, AZ), \quad (OA, AZ), \quad (OA, AC, CZ), \quad (OA, AB, BC, CZ) . \tag{23.3}$$

The set $\mathfrak{L}$ of links obtained from these paths is

$$\mathfrak{L} = \{ \{ 1, 4, 5 \}, \{ 1, 4, 5, 6 \}, \{ 2, 6 \}, \{ 2, 5 \}, \{ 2, 3, 4, 5 \} \} . \tag{23.4}$$

This set includes the minimal links, which are

$$\{ 1, 4, 5 \}, \quad \{ 2, 6 \}, \text{ and } \quad \{ 2, 5 \} . \tag{23.5}$$

[16] We do not write $\mathbf{a}' = \{ \Delta(u_1), \dots, \Delta(u_l) \}$ since to two different edges of the path may correspond the same component $\mathbf{e}_i$, which must be counted only once in $\mathbf{a}'$ so that this will be a set.

[17] Just as, in (23.1), the same component may be obtained from distinct arcs of the path μ, so in (23.2), the same link may be obtained from distinct paths of $\mathfrak{C}$; obviously it must be counted only once.

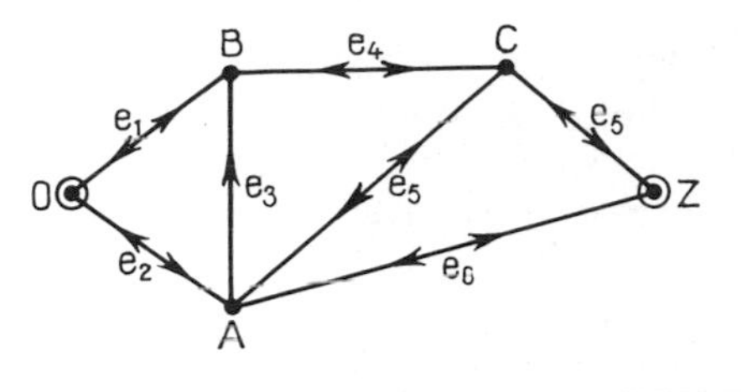

FIG. 23.1.

Note that the component $\{e_3\}$ is useless.

Theorem 23.I reduces the problem to the search for the elementary paths from O to Z in the graph G. For this, it is useful to consider the set $\mathcal{C}'$ composed of:

(1) the elements of $\mathcal{C}$ (elementary paths from O to Z),

(2) and the elementary paths with initial end O, with terminal end other than Z, and not passing through the vertex Z.

The elements of the set $\mathcal{C}'$ form a tree, that is, a 1-fold graph of a certain type[18] for which the set of vertices is the set $\mathcal{C}'$ and whose arcs connect the pairs of elementary paths such that

$$\mu_1 = (u_1, \ldots, u_l) \quad \text{and} \quad \mu_2 = (u_1, \ldots, u_l, u_{l+1}).$$

In other words, an elementary path is formed by adding one arc to a shorter elementary path. In this representation the vertex O of the graph G is considered as an elementary path; it constitutes the "root" of the tree.

Example. Figure 23.2 represents the tree of elementary paths (of the type

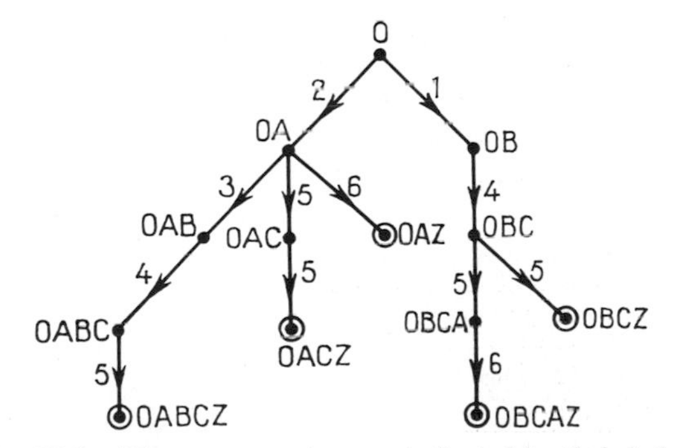

FIG. 23.2. The components e are indicated by their index i.

[18] A tree $G = (\mathbf{S}, \mathbf{U})$ with root $S_0 \in \mathbf{S}$ is a 1-fold connected graph such that: There exists no edge in $\mathbf{U}$ with terminal end S_0. For any $S_i \in \mathbf{S}$, $S_i \neq S_0$, there exists exactly one edge terminating at S_i.

defined above, that is, with initial end O and not passing through Z, the only vertex at which they may terminate) of the graph of Fig. 23.1.

In the case of a network having a small number of vertices, the tree of elementary paths may easily be constructed by hand, following a few simple precautions.

We go on to consider the example of Fig. 23.3, which is a little more complicated than that of Fig. 23.1.

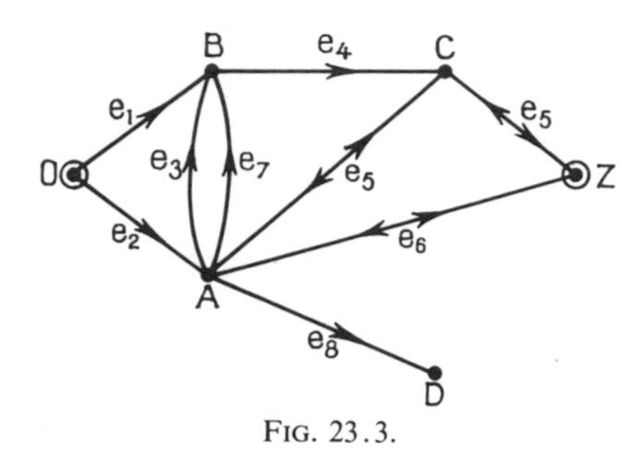

FIG. 23.3.

Example (Fig. 23.3). We arbitrarily choose an order among the arcs with given initial end. For example, we rank the terminal ends in the order A, B, C, D, Z; and in the case where there are several arcs with the same terminus, we rank in the order of increasing indices of components.

Starting at O, we construct a list of arcs with initial point O in the order above OA, OB. To each of these arcs will correspond a vertex of level 1 of the tree (Fig. 23.4), the level 0 being constituted by the vertex O. Beginning again at each of the vertices already constructed at the preceding level, we continue in the same manner, taking care at each stage to be sure that one does not return to a vertex already used. Whenever one reaches Z, one has obtained an elementary path from O to Z. One thus obtains in our example six elementary paths, to which correspond the following links (the indices of the components have been labeled on the arcs of the tree of Fig. 23.4 in order to facilitate locating the links):

$$(23.6)\quad \{e_2, e_3, e_4, e_5\},\quad \{e_2, e_4, e_5, e_7\},\quad \{e_2, e_5\},$$
$$\{e_2, e_6\},\quad \{e_1, e_4, e_5, e_6\},\quad \{e_1, e_4, e_5\}.$$

Among these links, three are minimal:

$$(23.7)\qquad \{e_2, e_5\},\quad \{e_2, e_6\},\ \text{and}\quad \{e_1, e_4, e_5\}.$$

The method described above is not convenient for use on a computer.

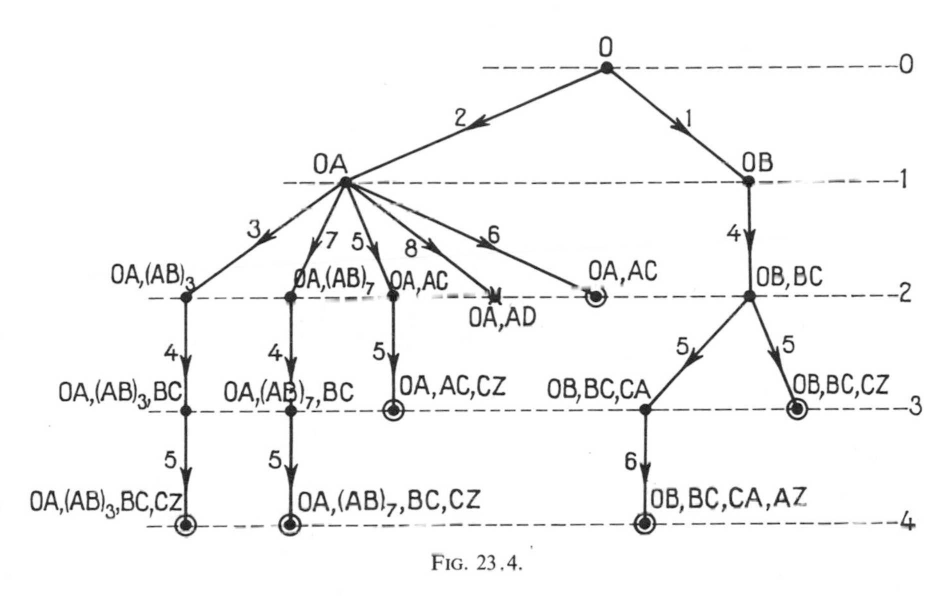

FIG. 23.4.

We take up below another method better suited for machine calculation which permits one to study complex networks.

Method of "Latin Composition" or "Concatenation."[19] We first show how to apply this method to enumerate, with neither redundance nor omission, all the elementary paths of a 1-fold graph; the case of a p-fold graph ($p > 1$) follows easily from this first case. After a first example concerning the search for all the elementary paths, a second example will show the application of the method in the case of a search for elementary paths between two given vertices, the case that interests us here.

With the aid of a particular type of matrix multiplication, it will be possible to enumerate successively, with neither redundance nor omission, all the elementary paths of length 1, 2, 3, ..., $n - 1$, where n is the number of vertices of the graph. The method will be introduced with two examples; as we shall see, it is very easy to understand.

Example 1. Consider the graph of Fig. 23.5 and construct a "latin matrix" in the following fashion: if a vertex X_i is joined to a vertex X_j by an arc (X_i, X_j), put $X_i X_j$ in square (X_i, X_j) of the matrix; if the pair is not an arc, indicate $\varnothing$ in that position of the matrix. Further, place $\varnothing$ in all squares where $X_i = X_j$ (the principal diagonal).

[19] See: A. Kaufmann, *Introduction à la Combinatorique*. Dunod, Paris, 1968; *Graphs, Dynamic Programming, and Finite Games*, Academic Press, New York, 1967; and A. Kaufmann and Y. Malgrange, *Rev. Fr. R. O.* **26**, 1963.

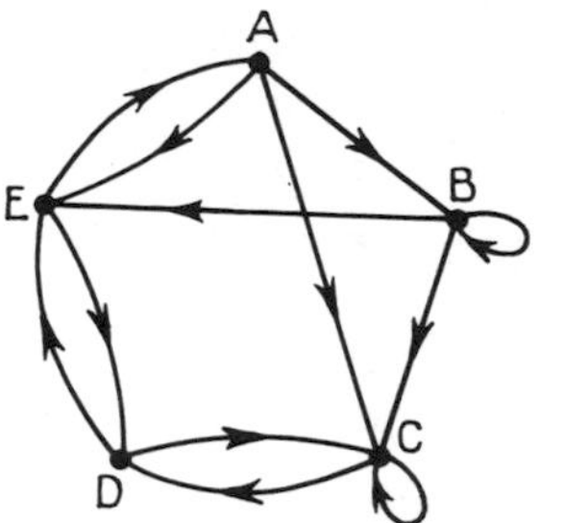

FIG. 23.5.

Thus the latin matrix relative to the graph of Fig. 23.5 is

(23.8) $[\mathcal{M}]^{(1)} =$

	A	B	C	D	E
A	∅	AB	AC	∅	AE
B	∅	∅	BC	∅	BE
C	∅	∅	∅	CD	∅
D	∅	∅	DC	∅	DE
E	EA	∅	∅	ED	∅

Now deduce from the latin matrix (23.8) another latin matrix in which the first letter has been suppressed; one has

(23.9) $[\tilde{\mathcal{M}}]^{(1)} =$

	A	B	C	D	E
A	∅	B	C	∅	E
B	∅	∅	C	∅	E
C	∅	∅	∅	D	∅
D	∅	∅	C	∅	E
E	A	∅	∅	D	∅

Hereafter, for clarity, any square containing ∅ will be left blank.

We now proceed to multiply $[\mathcal{M}]^{(1)}$ by $[\tilde{\mathcal{M}}]^{(1)}$ in a certain manner to form $[\mathcal{M}]^{(2)}$. Denote the elements of these matrices by $m_{i,j}^{(1)}$, $\tilde{m}_{i,j}^{(1)}$, $m_{i,j}^{(2)}$. For example, the element $m_{i,j}^{(1)}$ will be found in square (i, j), that is, in row i and column j of the matrix $[\mathcal{M}]^{(1)}$.

We obtain $m_{i,j}^{(2)}$ by the relation

$$m_{i,j}^{(2)} = \sum_k m_{i,k}^{(1)} \times \tilde{m}_{k,j}^{(1)}$$

where the symbols $\sum$ and $\times$ do not have their ordinary algebraic meaning, but rather have the following sense:

(a) $\quad m_{i,k}^{(1)} = \varnothing \quad \text{or} \quad \tilde{m}_{k,j}^{(1)} = \varnothing \quad \Rightarrow \quad m_{i,j}^{(1)} \times m_{k,j}^{(1)} = \varnothing;$

(b) if there is a letter common to $m_{i,k}^{(1)}$ and $\tilde{m}_{k,j}^{(1)}$, then their product is $\varnothing$;

(c) in any other case, the product $m_{i,k}^{(1)} \times \tilde{m}_{k,j}^{(1)}$ is obtained by combining the letters of the two terms;

(d) the symbol $\sum$ simply indicates the union of all the products $m_{i,k}^{(1)} \times \tilde{m}_{k,j}^{(1)}$.

We illustrate these rules with a small example: let $m_{i,k}^{(1)} = \{ AB, AC \}$ and $m_{k,j}^{(1)} = \{ D, C \}$; it then follows that

$$m_{i,k}^{(1)} \times m_{k,j}^{(1)} = \{ AB \times D, AB \times C, AC \times D, AC \times C \}$$

and we obtain

$$m_{i,k}^{(1)} \times \tilde{m}_{k,j}^{(1)} = \{ ABC, ABD, ACD \} .$$

Now carrying out the product of the matrices, (23.10) results:

(23.10)

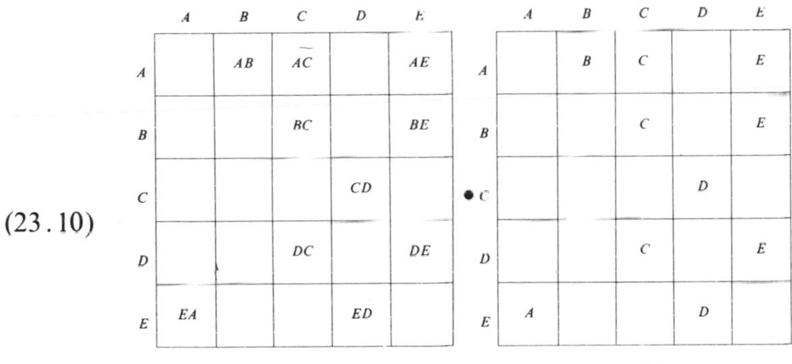

	A	B	C	D	E
A		AB	AC		AE
B			BC		BE
C				CD	
D			DC		DE
E	EA			ED	

$[\mathcal{M}]^{(1)}$

$\bullet$

	A	B	C	D	E
A		B	C		E
B			C		E
C				D	
D			C		E
E	A			D	

$[\tilde{\mathcal{M}}]^{(1)}$

$=$

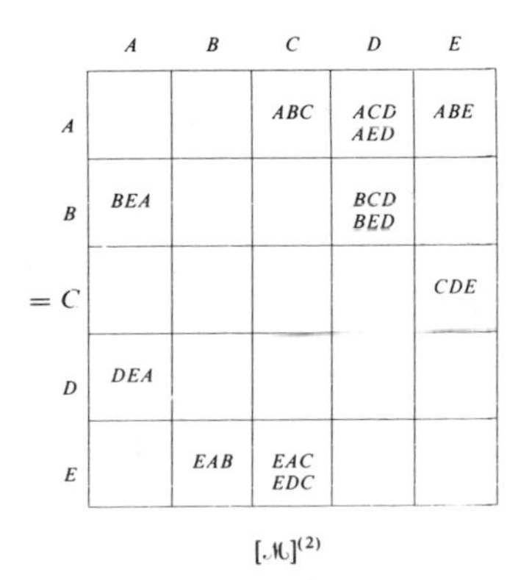

	A	B	C	D	E
A			ABC	ACD AED	ABE
B	BEA			BCD BED	
C					CDE
D	DEA				
E		EAB	EAC EDC		

$[\mathcal{M}]^{(2)}$

The latin matrix $[\mathcal{M}]^{(1)}$ gave all the elementary paths of length 1; the matrix $[\mathcal{M}]^{(2)}$ gives all the paths of length 2. We continue:

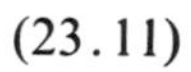

(23.11)

	A	B	C	D	E
A			ABC	ACD AED	ABE
B	BEA			BCD BED	
C					CDE
D	DEA				
E		EAB	EAC EDC		

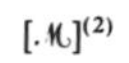

$[\mathcal{M}]^{(2)}$

$\bullet$

	A	B	C	D	E
A		B	C		E
B			C		E
C				D	
D			C		E
E	A			D	

$[\tilde{\mathcal{M}}]^{(1)}$

$=$

	A	B	C	D	E
A			AEDC	ABCD ABED	ACDE
B			BEAC BEDC		BCDE
C	CDEA				
D		DEAB	DEAC		
E			EABC	EACD	

$[\mathcal{M}]^{(3)}$

(23.12)

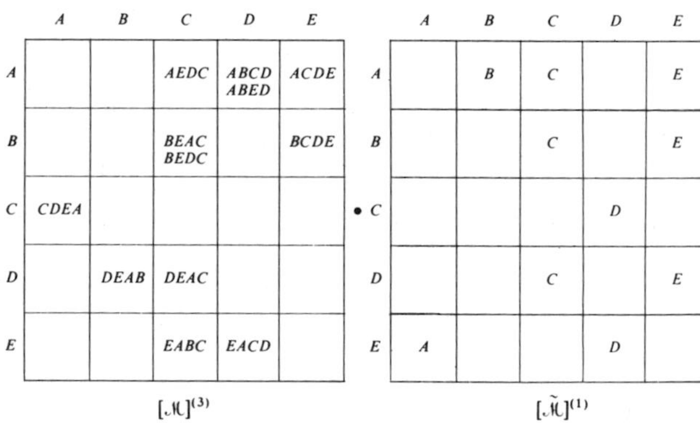

	A	B	C	D	E
A			AEDC	ABCD ABED	ACDE
B			BEAC BEDC		BCDE
C	CDEA				
D		DEAB	DEAC		
E			EABC	EACD	

$[\mathcal{M}]^{(3)}$

$\bullet$

	A	B	C	D	E
A		B	C		E
B			C		E
C				D	
D			C		E
E	A			D	

$[\tilde{\mathcal{M}}]^{(1)}$

$=$

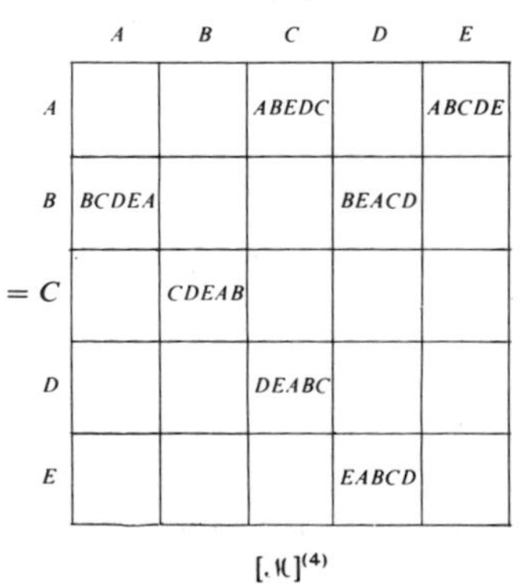

	A	B	C	D	E
A			ABEDC		ABCDE
B	BCDEA			BEACD	
C		CDEAB			
D			DEABC		
E				EABCD	

$[\mathcal{M}]^{(4)}$

(23.13)

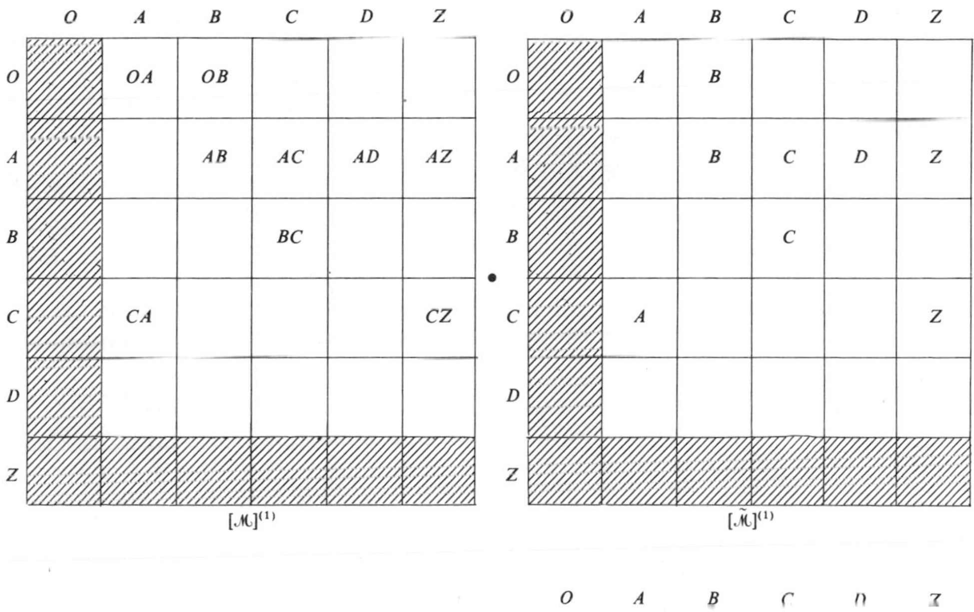

	O	A	B	C	D	Z
O		OA	OB			
A			AB	AC	AD	AZ
B				BC		
C		CA				CZ
D						
Z						

$[\mathcal{M}]^{(1)}$

•

	O	A	B	C	D	Z
O		A	B			
A			B	C	D	Z
B				C		
C		A				Z
D						
Z						

$[\tilde{\mathcal{M}}]^{(1)}$

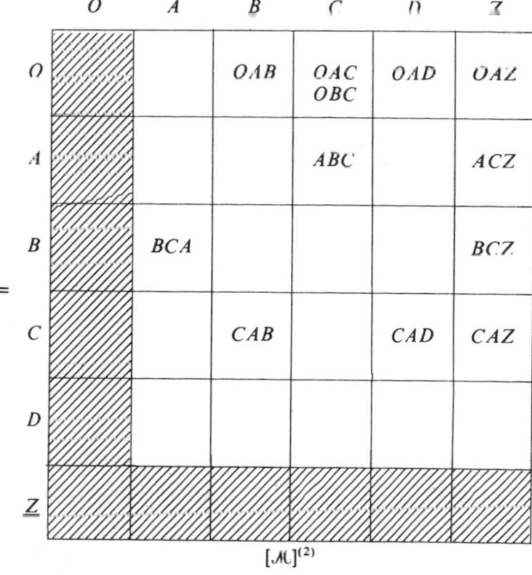

=

	O	A	B	C	D	Z
O			OAB	OAC OBC	OAD	OAZ
A				ABC		ACZ
B		BCA				BCZ
C			CAB		CAD	CAZ
D						
Z						

$[\mathcal{M}]^{(2)}$

(23.14)

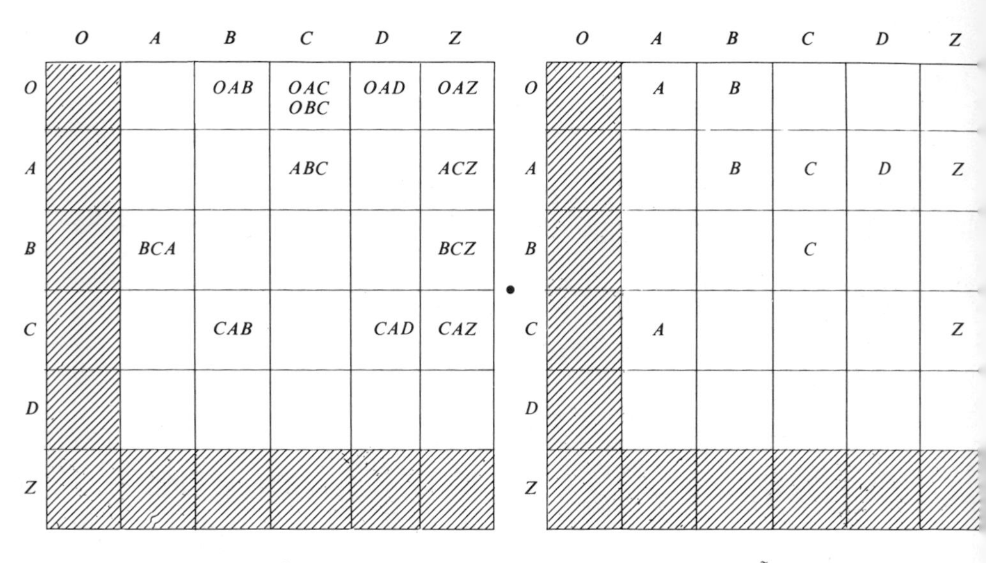

	O	A	B	C	D	Z
O			OAB	OAC OBC	OAD	OAZ
A				ABC		ACZ
B		BCA				BCZ
C			CAB		CAD	CAZ
D						
Z						

$[\mathcal{M}]^{(2)}$

•

	O	A	B	C	D	Z
O		A	B			
A			B	C	D	Z
B				C		
C		A				Z
D						
Z						

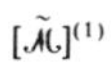

$[\tilde{\mathcal{M}}]^{(1)}$

=

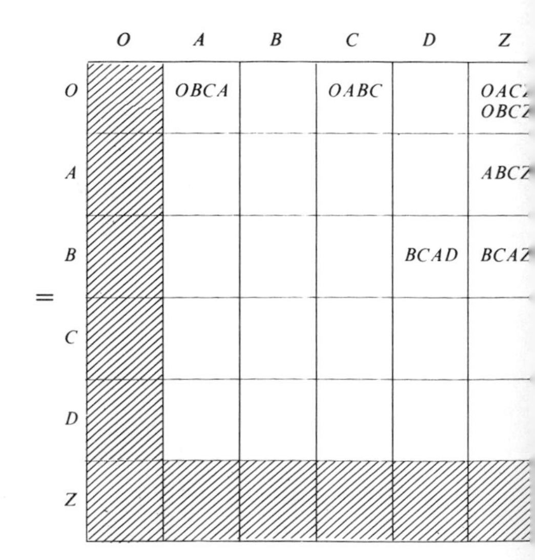

	O	A	B	C	D	Z
O		OBCA		OABC		OACZ OBCZ
A						ABCZ
B					BCAD	BCAZ
C						
D						
Z						

$[\mathcal{M}]^{(3)}$

(23.15)

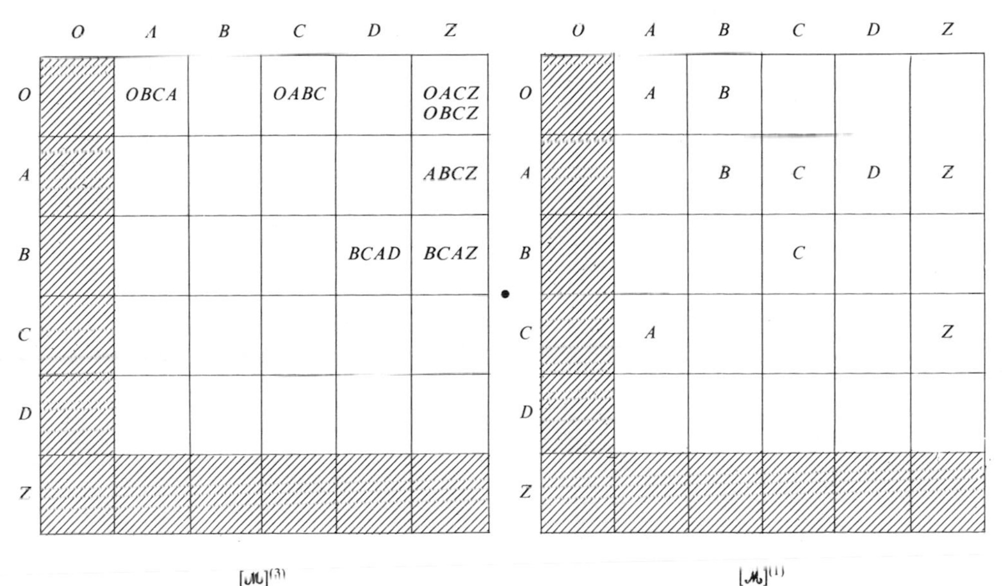

	O	A	B	C	D	Z
O		OBCA		OABC		OACZ OBCZ
A						ABCZ
B					BCAD	BCAZ
C						
D						
Z						

$[\mathcal{M}]^{(3)}$

•

	O	A	B	C	D	Z
O		A	B			
A			B	C	D	Z
B				C		
C		A				Z
D						
Z						

$[\mathcal{M}]^{(1)}$

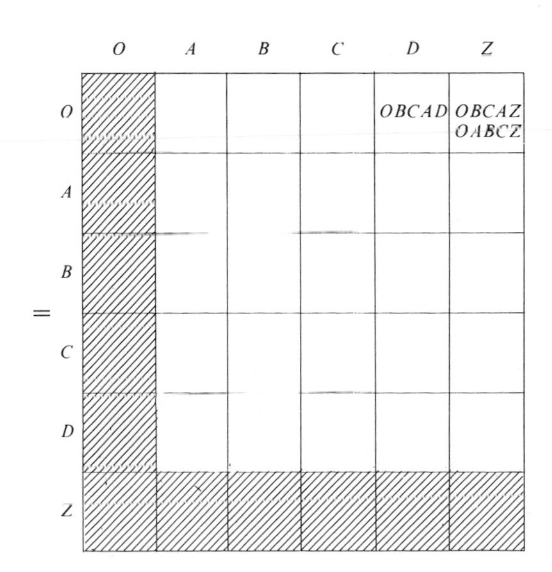

=

	O	A	B	C	D	Z
O					OBCAD	OBCAZ OABCZ
A						
B						
C						
D						
Z						

$[\mathcal{M}]^{(4)}$

The latin matrix $[\mathcal{M}]^{(3)}$ gives all the elementary paths of length 3; the matrix $[\mathcal{M}]^{(4)}$ gives all the elementary paths of length 4. It is unnecessary to go further; there are no elementary paths of length greater than 4 since the graph has five vertices.

The matrices $[\mathcal{M}]^{(i)}$, $i = 1, 2, 3, 4$, give, in order, the elementary paths of length $l = 1, 2, 3, 4$.

Note that this method of enumeration may be extended to all kinds of combinatoric concepts (see the first reference mentioned in footnote 19).

Example 2 (Fig. 23.6). This time we shall consider the enumeration of all elementary paths between two given vertices O and Z of a p-fold graph ($p = 2$ in this example). One reduces this to a 1-fold graph by not taking into account the two edges $(A, B)_{(3)}$ and $(A, B)_{(7)}$, but only a single edge (A, B). Once we have enumerated the links of the reliability network, we reintroduce the two arcs (thus two components) joining A and B.[20]

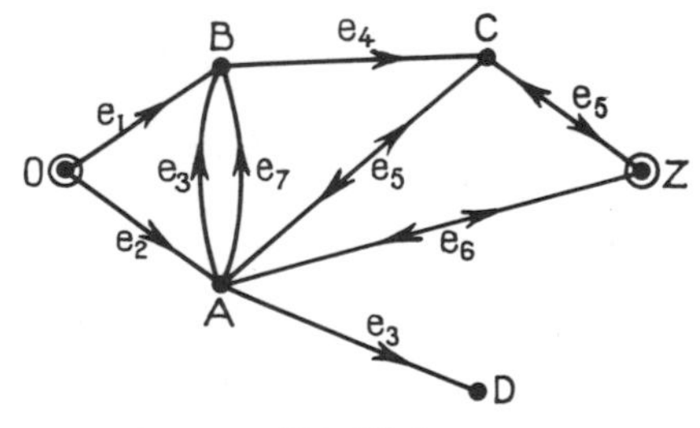

FIG. 23.6.

We give the matrices $[\mathcal{M}]^{(1)}$ and $[\tilde{\mathcal{M}}]^{(1)}$ in which, evidently, one may leave column O blank and row Z blank, since the paths ending at O or beginning at Z are not of interest in a reliability network.

The latin matrix $[\mathcal{M}]^{(2)}$ calculated in (23.13) already gives us an elementary path of length 2, namely (O, A, Z).

In the matrix $[\mathcal{M}]^{(3)}$ there are two new elementary paths (O, A, C, Z) and (O, B, C, Z).

The matrix $[\mathcal{M}]^{(4)}$ gives two paths of length 4: (O, B, C, A, Z) and (O, A, B, C, Z).

It is useless to go further to calculate $[\mathcal{M}]^{(5)}$; it is clear that there do not exist elementary paths of length 5 from O to Z.

Finally, from (23.13), (23.14), and (23.15) we have:

one elementary path of length 2: (O, A, Z);

(23.16) two elementary paths of length 3: (OA, C, Z) and (O, B, C, Z);

two elementary paths of length 4:

(O, B, C, A, Z) and (O, A, B, C, Z).

[20] One may also take account of multiple edges in latin multiplication.

To these elementary paths of the graph correspond the following links of the reliability network, following Fig. 23.6:

$$(O, A, Z) : \{ 2, 6 \} ,$$
$$(O, A, C, Z) : \{ 2, 5 \} ,$$
$$(O, B, C, Z) : \{ 1, 4, 5 \} , \qquad (23.17)$$
$$(O, B, C, A, Z) : \{ 1, 4, 5, 6 \} ,$$
$$(O, A, B, C, Z) : \{ 2, 3, 4, 5 \} \quad \text{and} \quad \{ 2, 7, 4, 5 \} .$$

From this, the three minimal links of the network are

$$\{ 2, 6 \} , \quad \{ 2, 5 \}, \quad \text{and} \quad \{ 1, 4, 5 \} . \qquad (23.18)$$

In fact, we have recovered (23.6) and (23.7), the chosen example being the same as that of Fig. 23.3.

A number of simplifications may be introduced in order to shorten the calculations.

Search for a Set of Cuts Including All Minimal Cuts of a Network. Having determined all minimal links, we establish the cuts by taking a component in each of the minimal links. The set of cuts thus obtained includes all minimal cuts (cf. Section 19, Theorem 19.IV).

Example. Consider again the example of Fig. 23.6. The minimal links are given by (23.18). The set of cuts may be obtained by proceeding as follows. We have noted previously that this set will include at most $2 \times 2 \times 3 = 12$ cuts. Let $\mathbf{A}_1 = \{ 2, 6 \}$, $\mathbf{A}_2 = \{ 2, 5 \}$, $\mathbf{A}_3 = \{ 1, 4, 5 \}$. Take 2 in $\mathbf{A}_1$, 2 in $\mathbf{A}_2$, 1 in $\mathbf{A}_3$; we thus form $\{ 2, 1 \}$. Take 2 in $\mathbf{A}_1$, 2 in $\mathbf{A}_2$, 4 in $\mathbf{A}_3$; we thus form $\{ 2, 4 \}$; and so on. We obtain

(23.19)

$$\{ \{ 2, 1 \}, \{ 2, 4 \}, \{ 2, 5 \}, \{ 2, 5, 1 \}, \{ 2, 5, 4 \}, \{ 6, 2, 1 \}, \{ 6, 2, 4 \}, \{ 6, 2, 5 \}, \{ 6, 5, 1 \}, \{ 6, 5, 4 \}, \{ 6, 5 \} \} .$$

The minimal cuts can then be extracted;

$$\{ \{ 2, 1 \}, \{ 2, 4 \}, \{ 2, 5 \}, \{ 5, 6 \} \} . \qquad (23.20)$$

Remark. The components e_3, e_7, and e_8, not belonging to any minimal link, are useless. The same conclusion may be obtained through considering the set of minimal cuts.

Search for Minimal Links and Minimal Cuts of a Monotone Structure Function.[21] We have seen, for Eq. (22.8) among others, how to determine the links and cuts of a structure function by enumerating all possible states of the components. We shall see here with an example how some simple "intuitions" permit one to avoid this enumeration, which is tedious when there is a large number of components.

Let the structure function be

(23.21)

$$\varphi(x_1, x_2, x_3, x_4, x_5) = x_1 x_4 + x_2 x_5 + x_1 x_3 x_5 + x_2 x_3 x_4 - x_1 x_2 x_3 x_5 - x_1 x_2 x_4 x_5 - x_1 x_3 x_4 x_5 - x_1 x_2 x_3 x_4 - x_2 x_3 x_4 x_5 + 2 x_1 x_2 x_3 x_4 x_5$$

where x_5 occurs in most of these monomials. Put $x_5 = 0$. Then we obtain

(23.22)

$$\varphi_1(x_1, x_2, x_3, x_4) = \varphi(x_1, x_2, x_3, x_4, 0) = x_1 x_4 + x_2 x_3 x_4 - x_1 x_2 x_3 x_4 .$$

With this simplified function, it is clear that

(23.23)

$$x_1 = x_4 = 1 \Rightarrow \varphi_1(x) = 1, \quad \text{giving the link} \quad \mathbf{a}_1 = \{1, 4\}$$

or

$$x_2 = x_3 = x_4 = 1 \Rightarrow \varphi_1(x) = 1, \quad \text{giving the link} \quad \mathbf{a}_2 = \{2, 3, 4\}$$

and it is clear that no other link of $\varphi_1(x)$ will be minimal; in fact, any link must give the value 1 to the first or second term of $\varphi_1(x)$, and therefore contains either $\mathbf{a}_1$ or $\mathbf{a}_2$.

Another component, x_2, also appears in a number of terms of $\varphi(x)$; now putting $x_2 = 0$ and $x_5 = 1$, we have

(23.24) $$\varphi_2(x_1, x_3, x_4) = \varphi(x_1, 0, x_3, x_4, 1) = x_1 x_4 + x_1 x_3 - x_1 x_3 x_4 ;$$

it is clear that

(23.25)

$$x_1 = x_4 = 1 \Rightarrow \varphi_2(x) = 1, \quad \text{link } \mathbf{a}_3 = \{1, 4, 5\};$$

$$x_1 = x_3 = 1 \Rightarrow \varphi_2(x) = 1, \quad \text{link } \mathbf{a}_4 = \{1, 3, 5\}.$$

Finally, putting $x_2 = x_5 = 1$, we obtain

(23.26) $$\varphi_3(x_1, x_3, x_4) = \varphi(x_1, 1, x_3, x_4, 1) \equiv 1$$

and

(23.27) $$x_2 = x_5 = 1 \Rightarrow \varphi(x) = 1, \quad \text{link } \mathbf{a}_5 = \{2, 5\}.$$

We have examined all possible cases since we have successively put $x_5 = 0$, then $x_5 = 1$, $x_2 = 0$, and $x_5 = x_2 = 1$. We have found five links of which one, $\mathbf{a}_3$, is not minimal since it contains the link $\mathbf{a}_1$.

[21] If a structure function is not monotone, the notion of a minimal link presents little interest.

There remain four minimal links:

$$(23.28)\qquad \begin{aligned} \mathbf{a}_1 &= \{ 1, 4 \}, \\ \mathbf{a}_2 &= \{ 2, 3, 4 \}, \\ \mathbf{a}_4 &= \{ 1, 3, 5 \}, \\ \mathbf{a}_5 &= \{ 2, 5 \}. \end{aligned}$$

We obtain the minimal cuts (and some nonminimal cuts) by taking a component in each minimal link:

(a) Take component 1 in $\mathbf{a}_1$; this component also occurs in $\mathbf{a}_4$. Take component 2 in $\mathbf{a}_2$; it also occurs in $\mathbf{a}_5$ and $\{ 1, 2 \}$ is a cut. Take component 3 in $\mathbf{a}_2$ and 5 in $\mathbf{a}_5$; we obtain the cut $\{ 1, 3, 5 \}$; it is evidently useless to take component 2 again in $\mathbf{a}_5$ since the minimal cut $\{ 1, 2 \}$ has already been identified. Take 4 in $\mathbf{a}_2$ and 5 in $\mathbf{a}_5$; we obtain the cut $\{ 1, 4, 5 \}$.

(b) Take component 4 in $\mathbf{a}_1$; this component also occurs in $\mathbf{a}_2$. Take 5 in $\mathbf{a}_4$ and in $\mathbf{a}_5$; we obtain the cut $\{ 4, 5 \}$. Take 1 in $\mathbf{a}_4$ and 2 in $\mathbf{a}_5$; we obtain $\{ 1, 2, 4 \}$. Take 3 in $\mathbf{a}_4$ and 2 in $\mathbf{a}_5$; we obtain $\{ 2, 3, 4 \}$.

Note that the cuts $\{ 1, 4, 5 \}$ and $\{ 1, 2, 4 \}$ are not minimal since they contain, respectively, the cuts $\{ 4, 5 \}$ and $\{ 1, 2 \}$.

There are thus four minimal cuts:

$$(23.29)\qquad \begin{aligned} \mathbf{b}_1 &= \{ 1, 2 \}, \\ \mathbf{b}_2 &= \{ 1, 3, 5 \}, \\ \mathbf{b}_3 &= \{ 4, 5 \}, \\ \mathbf{b}_4 &= \{ 2, 3, 4 \}. \end{aligned}$$

The reader may check that the network of Fig. 23.7 is equivalent to this structure function.

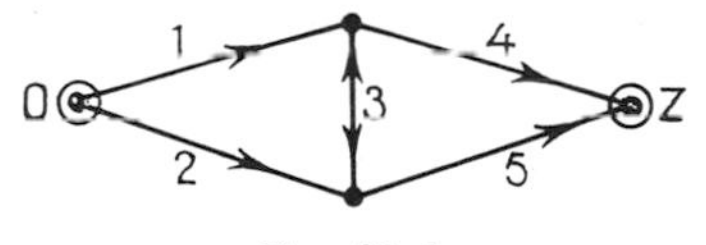

FIG. 23.7.

The method for searching for links and cuts of a structure function sketched above in fact uses the notion of linear composition of structures; this will be discussed in Section 26.

CHAPTER IV

STUDY OF THE RELIABILITY OF SYSTEMS

24 Introduction. Definitions and Hypotheses

We now begin the study of the reliability of complex systems with respect to that of their components; this study will be carried out with the tools developed in Chapter III (structure functions and reliability networks), and particular attention will be given to monotone structures.

The term *reliability* of a system may be extended, in a very general sense, to include the set of system characteristics that occur in its functioning more or less satisfactorily during a more or less lengthy time. More precisely, the reliability of a system is generally defined as the probability that it performs specified tasks, under specified use conditions, and during a specified time. This definition is, however, still too general to allow a mathematical study; to render it operative, it is necessary to indicate what the services that one expects from the system are, and what the conditions of use will be, which is evidently not easy in the framework of a theoretical study. Fortunately, the problem may be simplified by taking into account the hypotheses of Chapter I (Sections 1 and 2) and of Chapter III (Section 14), which we readopt in the present chapter and which are reviewed below:

(1) The system has only two possible states: either functioning well or failed.

(2) The system may be decomposed into r components in such a way that:

each component, at a given instant, either is in a good state or has failed; the state of the system depends only on the state of the set of its components.

(3) Each of the components e_i ($i = 1, 2, \ldots, r$) of the system has a random lifetime T_i; it is in a good state in the interval $(0, T_i)$ and has failed after the instant T_i. We designate the survival function[1] of the component as $v_i(t)$.

To these three hypotheses we add a fourth, which is unfortunately not always satisfied in practice, but which is difficult to do without in a general theory:

(4) The random variables T_i ($i = 1, 2, \ldots, r$) are independent.

In the definition given above for reliability, we supposed that the system performs the services that one expects of it if and only if it is in a good state (it is functioning). The conditions of use of the system do not appear explicitly, but only through the survival functions $v_i(t)$ of the components; for example, the components of a system subject to vibrations (as those mounted on any airplane or vehicle) will, in general, have a shorter lifetime than they would were the system stationary (cf. Section 2).

Formally, the reliability of a system may be defined in the following fashion:

Definition. *The reliability of a system satisfying hypotheses* (1)–(4) *is the probability that it is continually in a good state in the interval* $(0, t)$, *where* t *is a fixed time.*

The notion of reliability thus defined reduces easily to that of survival function as defined in Section 3. In fact, let T be the (random) instant at which the system falls in failure for the first time[2]; we shall call T the "lifetime" of the system. The survival function of the system is then

$$v(t) = \text{pr}\{T > t\}, \tag{24.1}$$

and the reliability of the system for the interval $(0, t)$ is no other than $v(t)$, the probability that the lifetime T of the system is greater than t.

In the case of a monotone structure, the irreversibility of the failure of the components entails the irreversibility of failure of the system. Then the event "the system is in a good state at time t" is the same as the event "the lifetime of the system is greater than t," and the reliability of the system is the probability that it will be in a good state at the instant t. It is so for the

[1] See Chapter 1, Section 3. If the component is not new at time 0, one would obviously use the corresponding survival function (see Section 7).

[2] We specify "for the first time" since, in the case of a nonmonotone structure, failure of a new component after the instant T might return the system to a good state.

components, according to hypothesis (3); we shall call "reliability" of a component e_i at time t the probability that it is in a good state at that instant, which is also the probability that it remains in a good state throughout the interval $(0, t)$.

In the case of a nonmonotone structure, the system may be in a good state at the instant t after having failed between 0 and t; the reliability of the system thus may be different from the probability that it is in a good state at time t.

In the present chapter we first establish the relation that exists between the probability that the system is in a good state at the instant t and the reliability of its components, a relation that we shall call the reliability function; and we shall study certain mathematical properties of reliability functions for monotone structures (Section 25). We then describe a very useful tool for the determination of the structure function or the reliability function of a system: composition; and we shall see how linear composition permits one to construct the set of monotone structures (Section 26); we then present a theorem of Moore and Shannon permitting the classification into three groups of the curves representative of reliability functions of monotone structures (Section 27). We end the study of the reliability of systems from a static point of view by presenting the notion of a system "monotone in probability" (Section 28). Finally, we examine the variation of the reliability as a function of time, that is, we pass to the study of the survival function of the system. We shall see that the failure rate of a system is not in general constant or monotone, even if that of its components is constant or monotone (Sections 29 and 30).

25 The Reliability Function

At a given instant t, a component e_i has a probability

$$p_i = v_i(t) \tag{25.1}$$

of being in a good state, and the complementary probability $1 - p_i$ of having failed. Let X_i be the random variable representing the state of the component e_i at the instant t, with the convention

$$\begin{aligned} X_i &= 1 \quad \text{if the component is in a good state} \\ &= 0 \quad \text{if the component has failed.} \end{aligned} \tag{25.2}$$

One therefore has

$$\begin{aligned} \text{pr}\,\{ X_i = 1 \} &= p_i\,, \\ \text{pr}\,\{ X_i = 0 \} &= 1 - p_i \end{aligned} \tag{25.3}$$

with

$$E(X_i) = 1 \times p_i + 0 \times (1 - p_i) = p_i\,. \tag{25.4}$$

Note that the independence of the lifetimes T_i of the components $(i = 1, 2, \ldots, r)$ entails the independence of the random variables X_i $(i = 1, 2, \ldots, r)$.

Let $(x) = (x_1, x_2, \ldots, x_r)$ be a possible value of the random r-tuple

$$(X) = (X_1, X_2, \ldots, X_r);$$

(x) is a "state of the set of components" (cf. Section 15) to which corresponds the state $\varphi(x)$ of the system. The function

$$\Phi = \varphi(X_1, X_2, \ldots, X_r) - \varphi(X) \tag{25.5}$$

is a certain function of the random variables $X_1, X_2, \ldots, X_r$. Therefore, this is also a random variable, which takes the value 1 if the system functions, and the value 0 when it has failed. The probability that $\Phi = 1$ is the probability that the system is in a good state at the instant considered. We shall call the random variable (25.5) the "random structure function" of the system.

The mathematical expectation of this random variable is

(25.6)

$$E(\Phi) = E[\varphi(X)] = 1 \times \text{pr}\{\Phi = 1\} + 0 \times \text{pr}\{\Phi = 0\} = \text{pr}\{\Phi = 1\}.$$

We shall call the function

$$h(p_1, p_2, \ldots, p_r) = E[\varphi(X)] \tag{25.7}$$

the "reliability function" of the system. It is indeed clear that $E[\varphi(X)]$ is a function of the reliabilities $p_1, p_2, \ldots, p_r$ of the components, which define the probability laws of the random variables $X_1, X_2, \ldots, X_r$. We proceed moreover immediately to obtain this reliability function. Indeed, suppose that the structure function of a system is in the simple form (Section 22), that is, in the form of a polynomial $\varphi_s(x)$ of first degree with respect to each of the variables $x_1, x_2, \ldots, x_r$. Then $\varphi(x)$ is a sum of terms of the form

$$k . X_{i_1} . X_{i_2} . \ldots . X_{i_l} \qquad (l \leqslant r).$$

Since the random variables $X_{i_1}, \ldots, X_{i_l}$ are independent, we have

$$\begin{aligned} E(k . X_{i_1} . \ldots . X_{i_l}) &= k . E(X_{i_1}) . E(X_{i_2}) . \ldots . E(X_{i_l}) \\ &= k . p_{i_1} . p_{i_2} . \ldots . p_{i_l}. \end{aligned} \tag{25.8}$$

It then follows that

$$h(p) = h(p_1, p_2, \ldots, p_r) \equiv \varphi_s(p_1, p_2, \ldots, p_r) = \varphi_s(p) \tag{25.9}$$

where

$$(p) = (p_1, p_2, \ldots, p_r).$$

In other words, the reliability function is obtained simply by replacing the the variables $x_1, x_2, \ldots, x_r$ by $p_1, p_2, \ldots, p_r$ in the structure function of the system *expressed in simple form.*

Examples. (1) Consider again the example of Fig. 23.1, which is reproduced as Fig. 25.1 for convenience. We have seen in (23.5) that the system represented by this network has minimal links

$$\{2, 5\}, \quad \{2, 6\}, \quad \{1, 4, 5\}. \tag{25.10}$$

According to the general formula (22.1), the survival function is then

$$\varphi(x) = 1 - (1 - x_2 x_5)(1 - x_2 x_6)(1 - x_1 x_4 x_5). \tag{25.11}$$

Developing and simplifying this expression we obtain

$$\varphi_s(x) = x_2 x_5 + x_2 x_6 + x_1 x_4 x_5 - x_2 x_5 x_6 - x_1 x_2 x_4 x_5. \tag{25.12}$$

In order to obtain the reliability function it suffices to replace, purely and simply, x_i by p_i, $i = 1, 2, 3, 4, 5$; we thus have

$$h(p) = p_2 p_5 + p_2 p_6 + p_1 p_4 p_5 - p_2 p_5 p_6 - p_1 p_2 p_4 p_5 \tag{25.13}$$

where p_i is the reliability of component e_i, $i = 1, 2, 3, 4, 5$.

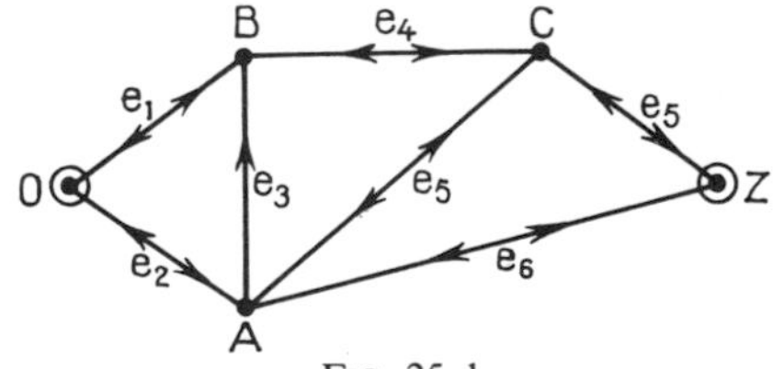

FIG. 25.1.

Note that the reliability of the useless component e_3 does not occur in the reliability function $h(p)$ of the system, as one would expect.

(2) The nonmonotone structure function (22.17), which we rewrite here as

$$\varphi(x_1, x_2, x_3) = (1 - x_1 x_2 x_3)(1 - x_2 x_3) x_1 x_3 \tag{25.14}$$

has the simple form (22.19):

$$\varphi_s(x_1, x_2, x_3) = x_1 x_3 - x_1 x_2 x_3. \tag{25.15}$$

Its reliability function is therefore

$$h(p_1, p_2, p_3) = p_1 p_3 - p_1 p_2 p_3 = p_1 p_3(1 - p_2). \tag{25.16}$$

The properties that allow one to pass from (25.14) to (25.15), that is, having $x_i^k = x_i$ for all k, is evidently not valid for the reliabilities p_i; hence, replacing the x_i by the p_i in (25.14) would result in a function that would be different from (25.16) and that would not be the reliability function.

Other Expressions of the Reliability Function. The states of the set of components being disjoint events in the sense of probability theory, we obtain the reliability function by adding the probabilities of all the links. An arbitrary r-tuple $(x) = (x_1, x_2, \ldots, x_r)$ has probability equal to $\alpha_1(x) \cdots \alpha_i(x) \cdots \alpha_r(x)$ where

$$\begin{aligned} \alpha_i(x_i) &= p_i \qquad \text{if} \quad x_i = 1\,, \\ &= 1 - p_i \quad \text{if} \quad x_i = 0\,. \end{aligned} \tag{25.17}$$

We may then write

$$h(p) = \sum_{x \in \mathfrak{C}} \varphi(x_1, \ldots, x_r)\, \alpha_1(x_1) \ldots \alpha_i(x_i) \ldots \alpha_r(x_r)\,, \tag{25.18}$$

where $\mathfrak{C}$ is the set of 2^r r-tuples. The r-tuples for which $\varphi = 0$ gives a zero term; those for which $\varphi = 1$, corresponding to a link (cf. Section 17), give a term equal to the probability of the r-tuple being considered. This expression may also be obtained from (22.18).

Example. The table of values of the structure function (25.14), which is given in Fig. 22.8, shows that the only state of the set of components giving φ the value 1 is the state (1, 0, 1), to which corresponds the unique link $\{ e_1, e_3 \}$ of the structure function. Formula (25.18) then carries only one nonzero term

$$h(p) = p_1(1 - p_2)\, p_3\,.$$

Indeed we have recovered (25.16).

Case Where All Components Have the Same Reliability. If all the components have the same reliability p, the probability of observing a fixed state corresponding to k components in good states with $r - k$ defective is given by a binomial law:

$$\pi(k) = p^k(1 - p)^{r-k}\,, \qquad k = 0, 1, 2, \ldots, r\,. \tag{25.19}$$

Let A_k be the number of links having k components; we then obtain a simple expression for the reliability function

$$h(p) = \sum_{k=0}^{r} A_k\, p^k(1 - p)^{r-k}\,. \tag{25.20}$$

Similarly, let B_k be the number of cuts having $r - k$ components; we thus obtain

$$1 - h(p) = \sum_{k=0}^{r} B_k\, p^{r-k}(1 - p)^k\,. \tag{25.21}$$

Example. Consider again the network of Fig. 25.1. This network has three minimal links, listed in (25.10); we obtain all links by considering the subsets of components including a minimal link. We thus have:

0	links having 0 components:	$A_0 = 0;$
0	links having 1 component:	$A_1 = 0;$
2	links having 2 components (these are { 2, 5 } and { 2, 6 }):	$A_2 = 2;$
8	links having 3 components (these are { 2, 5, 1 }, { 2, 5, 3 }, { 2, 5, 4 }, { 2, 5, 6 }, { 2, 1, 6 } { 2, 3, 6 }, { 2, 4, 6 }, { 1, 4, 5 }):	$A_3 = 8;$
11	links having 4 components (there are $\binom{6}{4} = 15$ possible states, but the complements of the four cuts { 2, 1 }, { 2, 4 }, { 2, 5 }, { 6, 5 } are not links):	$A_4 = 11;$
6	links having 5 components (there are $\binom{6}{5} = 6$ possible states, and no cut with 1 component):	$A_5 = 6;$
1	link having 6 components (this is { 1, 2, 3, 4, 5, 6 }):	$A_6 = 1.$

We finally obtain the reliability function

(25.22)

$$h(p) = 2\,p^2(1-p)^4 + 8\,p^3(1-p)^3 + 11\,p^4(1-p)^2 + 6\,p^5(1-p) + p^6 .$$

By expanding this expression we obtain

(25.23)

$$h(p) = 2\,p^2 - p^4 = p^2(2 - p^2) .$$

The same result may indeed be obtained by putting $p_1 = p_2 = p_3 = p_4 = p_5 = p_6 = p$ into expression (25.13).

Properties of the Reliability Function of a Monotone Structure. We now present a sequence of theorems stating various important properties of the reliability function of a monotone structure.

Theorem 25.I. *The reliability function $h(p)$ of a monotone structure is monotone, that is,*

(25.24)

$$(q) \succcurlyeq (p) \quad \Rightarrow \quad h(q) \geqslant h(p) .$$

To prove this theorem we show that, for any vector $p = (p_1, p_2, \dots, p_r)$ such that $0 \leqslant p_i \leqslant 1$, $i = 1, \dots, r$, the various partials $\partial h(p)/\partial p_i$ are nonnegative. Consider expression (25.18) of the reliability function, and let $\mathfrak{C}'_i$ be the set of 2^{r-1} $(r-1)$-tuples

$$(x') = (x_1, \dots, x_{i-1}, x_{i+1}, \dots, x_r)\,. \tag{25.25}$$

We obtain the 2^r r-tuples (x) by completing each of the x' either by $x_i = 1$ or by $x_i = 0$. We thus have

(25.26)

$$h(p) = \sum_{x' \in \mathfrak{C}'_i} \varphi(x_1, \dots, x_{i-1}, 1, x_{i+1}, \dots, x_r)\, \alpha_1(x_1) \dots \alpha_{i-1}(x_{i-1}) . p_i \dots \alpha_r(x_r)$$
$$+ \sum_{x' \in \mathfrak{C}'_i} \varphi(x_1, \dots, x_{i-1}, 0, x_{i+1}, \dots, x_r)\, \alpha_1(x_1) \dots \alpha_{i-1}(x_{i-1}) . (1-p_i) \dots \alpha_r(x_r),$$

from which

(25.27)

$$\frac{\partial h(p)}{\partial p_i} = \sum_{x' \in \mathfrak{C}'_i} [\varphi(x_1, \dots, x_{i-1}, 1, x_{i+1}, \dots, x_r) - \varphi(x_1, \dots, x_{i-1}, 0, x_{i+1}, \dots, x_r)]$$
$$\times\, \alpha_1(x_1) \dots \alpha_{i-1}(x_{i-1})\, \alpha_{i+1}(x_{i+1}) \dots \alpha_r(x_r)\,.$$

Since the structure function φ is monotone, all the terms between square brackets are nonnegative (they are equal to 0 or to 1), and the partial derivative is nonnegative.

Theorem 25.II [19]. *Let $\varphi(x)$ be a monotone structure function; $h(p)$ its reliability function; $\mathbf{b}_1, \mathbf{b}_2, \dots, \mathbf{b}_n$ the minimal cuts of this structure; and $\mathbf{a}_1, \mathbf{a}_2, \dots, \mathbf{a}_m$ the minimal links. Then*

$$\prod_{k=1}^{n} \left[1 - \prod_{e_i \in \mathbf{b}_k} (1 - p_i)\right] \leqslant h(p) \leqslant 1 - \prod_{j=1}^{m} \left[1 - \prod_{e_i \in \mathbf{a}_j} p_i\right]. \tag{25.28}$$

Note that the first member of the inequality above will be equal to the reliability function $h(p)$ if all the minimal cuts are pairwise disjoint. Similarly, $h(p)$ will be equal to the third member if no two of the minimal links have a component in common.

Example. Consider the monotone structure function

$$\varphi(x) = x_1\, x_2 + x_1\, x_3 - x_1\, x_2\, x_3 \tag{25.29}$$

whose minimal links are $\mathbf{a}_1 = \{ 1, 2 \}$ and $\mathbf{a}_2 = \{ 1, 3 \}$ and whose minimal cuts are $\mathbf{b}_1 = \{ 1 \}$ and $\mathbf{b}_2 = \{ 2, 3 \}$. The inequality above becomes

(25.30)

$$[1-(1-p_1)]\,[1-(1-p_2)\,(1-p_3)] \leqslant h(p) \leqslant 1-(1-p_1\, p_2)\,(1-p_1\, p_3)$$

and we have

$$h(p) = p_1 p_2 + p_1 p_3 - p_1 p_2 p_3 \tag{25.31}$$

since (25.29) is in simple form. Since the cuts are disjoint, the first member is equal to $h(p)$, but the third member is equal to $h(p) + (1 - p_1)p_1 p_2 p_3$.

The two inequalities that follow involve the partial derivatives of the function $h(p)$. The first may be deduced [5] from the Schwarz inequality; the second is a generalized form of a result of Moore and Shannon (see Birnbaum *et al.* [8]).

Theorem 25.III. *The reliability function $h(p)$ of a monotone structure satisfies the following inequalities:*

$$h(p)\,[1 - h(p)] \sum_{i=1}^{r} p_i(1 - p_i) \geqslant \left[\sum_{i=1}^{r} p_i(1 - p_i) \frac{\partial h(p)}{\partial p_i} \right]^2, \tag{25.32}$$

$$h(p)\,[1 - h(p)] \leqslant \sum_{i=1}^{r} p_i(1 - p_i) \frac{\partial h(p)}{\partial p_i}. \tag{25.33}$$

An abbreviated form of the Moore–Shannon inequality (25.33) is given by the expression

$$\mathrm{Cov}\,[S(X), \varphi(X)] \geqslant \mathrm{Var}\,\varphi(X) \tag{25.34}$$

where

$$S(X) = X_1 + X_2 + \cdots + X_r \tag{25.35}$$

and Cov means "covariance of," Var means "variance of."

Theorem 25.IV. *Put*

$$\bar{a} = E[S(X) \mid \varphi(X) = 1], \tag{25.36}$$

$$\bar{b} = E[r - S(X) \mid \varphi(x) = 0], \tag{25.37}$$

where $S(X)$ is defined by (25.35), *$\bar{a}$ is the mean number of components in links, and $\bar{b}$ the mean number of components in cuts. One then has*

$$\bar{a} + \bar{b} \geqslant n + 1, \tag{25.38}$$

where n is the order of the monotone structure function $\varphi(x)$.[3]

[3] See Section 21, p. 86.

Theorem 25.V. *Let λ be the length[4] of a monotone structure and μ its width; if the structure is made up of r components each having the same reliability p, then one has*

$$p^{\lambda} \leqslant h(p) \leqslant 1 - (1 - p)^{\mu} . \tag{25.39}$$

Indeed, the good functioning of the λ components of the smallest link suffices to assure that the system will function, and the failure of the smallest cut suffices to bring the system to failure.

26 Composition of Structures

Definition. Composition operation for structure functions. *Let $\gamma(x)$ be a structure function where $(x) = (x_1, x_2, \dots, x_m)$, $x_1, x_2, \dots, x_m$ constituting a family of structure functions*

$$x_1 = x_1(u_1, u_2, \dots, u_{k_1}),\ x_2 = x_2(v_1, v_2, \dots, v_{k_2}),\ \dots,\ x_m = x_m(z_1, z_2, \dots, z_{k_m});$$

then

$$\varphi(u, v, \dots, z) = \gamma(x_1(u), x_2(v), \dots, x_m(z)) \tag{26.1}$$

is called the composition of $x_1, x_2, \dots, x_m$ in γ, and $x_1(u), x_2(v), \dots, x_m(z)$ are called "substructures" of $\varphi(u, v, \dots, z)$.

If $(u), (v), \dots, (z)$ are independent r-tuples (that is, if no two of these r-tuples have a component in common), then $x_1(u), x_2(v), \dots, x_m(z)$ are called "modules" of φ; in this case, if γ does not have useless components (if it is of order m), the order of φ is the sum of the orders of $x_1, x_2, \dots, x_m$.

Example. Let

(26.2)
(26.3)
$$x_1 = 1 - (1 - u_1)(1 - u_2), \quad x_2 = 1 - (1 - v_1)(1 - v_2)$$

and

$$\gamma(x_1, x_2) = x_1 \cdot x_2 . \tag{26.4}$$

Then

$$\begin{aligned} \varphi(u, v) &= [1 - (1 - u_1)(1 - u_2)][1 - (1 - v_1)(1 - v_2)] \\ &= (u_1 + u_2 - u_1 u_2)(v_1 + v_2 - v_1 v_2) . \end{aligned} \tag{26.5}$$

Definition. Composition operation for reliability networks. *Let $\mathfrak{Q}$ be a reliability network of order m (with m useful components) and $r_1, r_2, \dots, r_m$ be*

[4] The length and width of a structure have been defined in Section 17, p. 66.

a family of reliability networks with respective orders k_1, k_2, k_m. If each arc corresponding to component x_i in the network $\mathfrak{Q}$ is replaced by the network r_i, we obtain a network $\mathfrak{R}$ called the "composition of $r_1, r_2, \ldots, r_m$ in $\mathfrak{Q}$," and we write

$$\mathfrak{R} = \mathfrak{Q}(r_1, r_2, \ldots, r_m) . \tag{26.6}$$

We shall call $r_1, r_2, \ldots, r_m$ "subnetworks" of $\mathfrak{R}$. If $r_1, r_2, \ldots, r_m$ are independent, they are also called "modules" of $\mathfrak{R}$.

Example (Figs. 26.1 and 26.2). In Fig. 26.1 we have represented a network $\mathfrak{Q}$ that contains three subnetworks r_1, r_2, and r_3 which are not independent. For example, r_1 and r_2 contain a common component e_1.

In Fig. 26.2 we have represented a network $\mathfrak{Q}$ that contains four independent subnetworks or modules. None of these modules contains a component existing in another module.

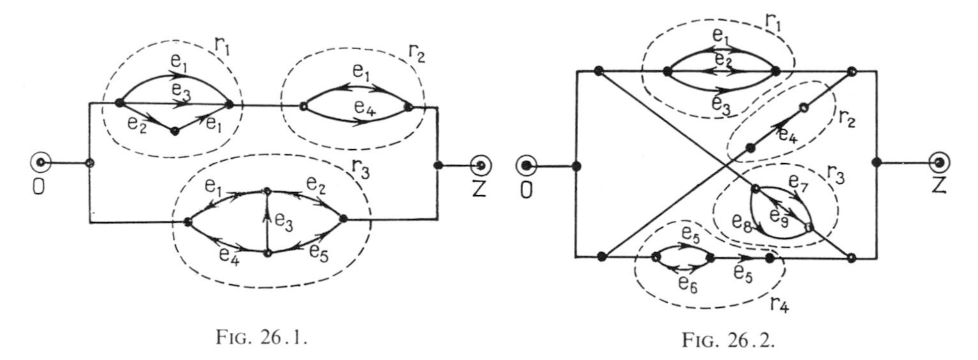

FIG. 26.1. FIG. 26.2.

Theorem 26.I. *If the network $\mathfrak{Q}$ is equivalent to the structure function $\mathfrak{R}$, and the networks $r_1, r_2, \ldots, r_m$ are, respectively, equivalent to structure functions $\mu_1, \mu_2, \ldots, \mu_m$, then the network $\mathfrak{R} = \mathfrak{Q}(r_1, r_2, \ldots, r_m)$ is equivalent to the function*

$$\varphi(u, v, \ldots, z) = \varphi[\mu_1(u), \mu_2(v), \ldots, \mu_m(z)] . \tag{26.7}$$

This theorem may easily be checked by enumerating the links of $\mathfrak{R}$ and $\mathfrak{Q}$.

Composition of Reliability Functions. Let

$$\varphi(u, v, \ldots, z) = \gamma[x_1(u), x_2(v), \ldots, x_n(z)]$$

be a structure function having $x_1(u)$, $x_2(v)$, $\ldots$, $x_n(z)$ for modules (that is, $u, v, \ldots, z$ are independent r-tuples), and let $g(p), f_1(q^1), f_2(q^2), \ldots, f_n(q^n)$ be the reliability functions corresponding, respectively, to the structures γ, $x_1, x_2, \ldots, x_n$. We then have the following theorem.

Theorem 26.II. *The reliability function of the structure* φ *is*

$$h(q^1, q^2, \ldots, q^n) = g[f_1(q^1), f_2(q^2), \ldots, f_n(q^n)] . \tag{26.8}$$

This theorem may easily be deduced from relations (25.6) and (25.7); $f_i(q^i)$ is by definition the mathematical expectation of X_i, but it is also the probability p_i that $X_i = 1$.

The examples that follow show the simplifications brought about by composition for the calculation of reliability functions.

Use of Composition for the Calculation of Structure Functions and Reliability Functions.

Example 1. Consider the network $\mathfrak{R}$ given in Fig. 26.3. This network may be decomposed into three subnetworks of which two are identical (Fig. 26.4). The subnetworks are arranged as indicated in Fig. 26.5.

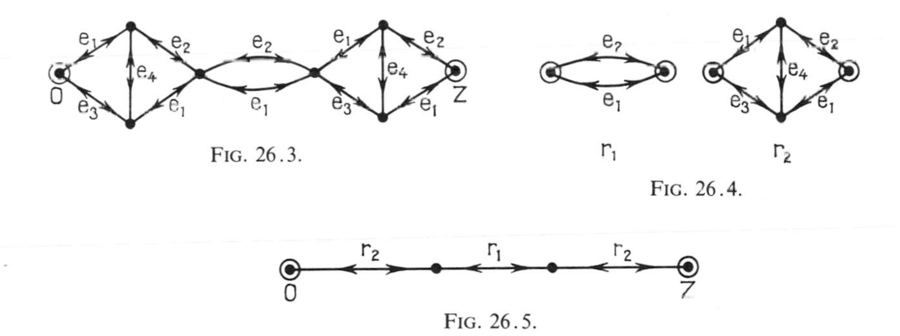

FIG. 26.3.

FIG. 26.4.

FIG. 26.5.

The structure function corresponding to the network $\mathfrak{Q}$ of Fig. 26.5 is

$$\gamma = \mu_1 . (\mu_2)^2 = \mu_1 . \mu_2 , \tag{26.9}$$

with

$$\mu_1 = 1 - (1 - x_1)(1 - x_2) \tag{26.10}$$

and

$$\mu_2 = 1 - (1 - x_1 x_2)(1 - x_1 x_3)(1 - x_1 x_4)(1 - x_2 x_3 x_4) \tag{26.11}$$

since the minimal links of r_1 are { 1 } and { 2 }, those of r_2 are { 1, 2 }, { 1, 3 }, { 1, 4 }, and { 2, 3, 4 }.

Putting (26.10) and (26.11) into (26.9), we obtain after reduction

(26.12)

$$\varphi(x_1, x_2, x_3, x_4) = x_1 x_2 + x_1 x_3 + x_1 x_4 - x_1 x_2 x_3 - x_1 x_2 x_4 - x_1 x_3 x_4 + x_2 x_3 x_4 .$$

If we further calculate the expanded form of μ_2, we find that

$$\mu_2(x_1, x_2, x_3, x_4) = \varphi(x_1, x_2, x_3, x_4) .$$

The fact that finally the network $\mathfrak{R}$ reduced to r_2 might have been revealed a priori. In fact, each minimal link of r_2 includes a minimal link of r_1 and therefore constitutes a link of $\mathfrak{R}$.

The reliability function corresponding to (26.12) is

(26.13)

$$h(p_1, p_2, p_3, p_4) = p_1 p_2 + p_1 p_3 + p_1 p_4 - p_1 p_2 p_3 - p_1 p_2 p_4 - p_1 p_3 p_4 + p_2 p_3 p_4 .$$

Suppose that all the components have the same reliability p, then we have

(26.14) $$h(p) = 3 p^2 - 2 p^3 .$$

Remark. In the example of Fig. 26.3, the subnetworks are not independent and thus do not form modules; therefore we may not calculate the reliability function by passing directly from $\gamma = \mu_1 \mu_2$ to this function; it is necessary to take into account the common components.

Example 2 (Figs. 26.6–26.8). This time the network $\mathfrak{R}'$ is formed of three subnetworks that are modules. In this case the structure function is

(26.15) $$\gamma' = \mu_1 \cdot \mu_2 \cdot \mu_3 .$$

Suppose that all the components have the same reliability p. For the network r_1', we have

(26.16) $$h_1(p) = 3 p^2 - 2 p^3 .$$

For the network r_2,

(26.17) $$h_2(p) = 2 p - p^2 .$$

For the network r_3,

(26.18) $$h_3(p) = 3 p^2 - 2 p^3 .$$

Finally, for the network $\mathfrak{R}'$,

$$
\begin{aligned}
h(p) &= h_1(p) . h_2(p) . h_3(p) \\
&= (2\,p - p^2)\,(3\,p^2 - 2\,p^3)^2 .
\end{aligned}
\tag{26.19}
$$

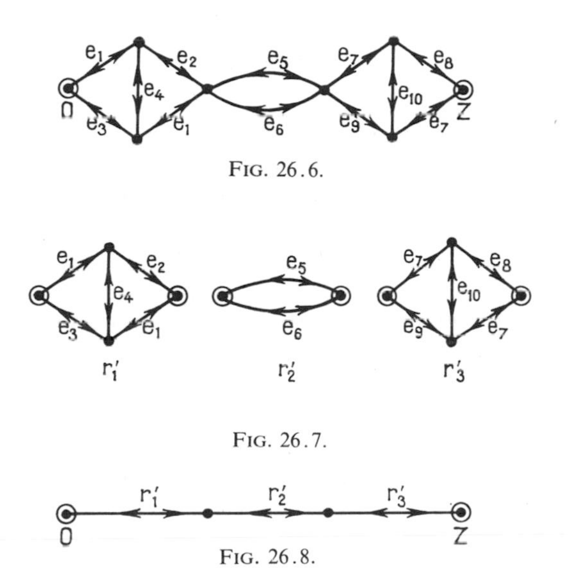

FIG. 26.6.

FIG. 26.7.

FIG. 26.8.

Remark. Of course, the decomposition of a network into subnetworks may be carried out in a number of ways; in general one seeks to give a series or parallel structure to the network $\mathfrak{R}$.

Linear Composition of Structure Functions. The linear composition of two structure functions of order at most n, $\varphi_1(x_1, x_2, \ldots, x_n)$ and $\varphi_2(x_1, x_2, \ldots, x_n)$, is defined as the structure of order at most $n + 1$, $\varphi(x_1, x_2, \ldots, x_n, x_{n+1})$, defined by the relation

$$
\begin{aligned}
\varphi(x_1, x_2, \ldots, x_n, x_{n+1}) &= x_{n+1}\,\varphi_1(x_1, x_2, \ldots, x_n) \\
&\quad + (1 - x_{n+1})\,\varphi_2(x_1, x_2, \ldots, x_n) .
\end{aligned}
\tag{26.20}
$$

Thus φ is obtained by composition of the structures

$$
\mu_1 = x_{n+1}\,, \quad \mu_2 = \varphi_1(x_1, x_2, \ldots, x_n)\,, \quad \mu_3 = \varphi_2(x_1, x_2, \ldots, x_n)\,,
\tag{26.21}
$$

in the structure

$$
\gamma = \mu_1\,\mu_2 + (1 - \mu_1)\,\mu_3 .
\tag{26.22}
$$

Important Remark. The structure γ given by (26.22) is not, in general, monotone; one may convince oneself of this through the table in Fig. 26.9 by comparing rows 6 and 2. Thus linear composition does not have a simple geometric correspondence in the domain of reliability networks. We shall see, however, that linear composition preserves the monotonicity of structures φ_1 and φ_2 if $\varphi_1 \geqslant \varphi_2$ since then γ is monotone.

	μ_1	μ_2	μ_3	γ
(1)	0	0	0	0
(2)	0	0	1	1
(3)	0	1	0	0
(4)	0	1	1	1
(5)	1	0	0	0
(6)	1	0	1	0
(7)	1	1	0	1
(8)	1	1	1	1

FIG. 26.9.

Theorem 26.III. *Any structure function of order n is a linear composition of two structure functions of order at most equal to $n - 1$.*

Indeed, one may write

$$(26.23)\quad \varphi(x_1, x_2, \ldots, x_{n-1}, x_n) = x_n\, \varphi(x_1, x_2, \ldots, x_{n-1}, 1) + (1 - x_n)\, \varphi(x_1, x_2, \ldots, x_{n-1}, 0)\,.$$

Theorem 26.IV. *Any structure function $\varphi(x_1, x_2, \ldots, x_n)$ may be written in the form*

$$(26.24)\qquad \varphi(x_1, x_2, \ldots, x_n) = \sum_{\xi \in \mathfrak{C}} \varphi(\xi) \prod_{j=1}^{n} x_j^{\xi_j}(1 - x_j)^{1-\xi_j}\,,$$

where $\mathfrak{C}$ designates the set of 2^n states $\xi = (\xi_1, \xi_2, \ldots, \xi_n)$ of the set of components.

This expression may be obtained by repeating the decomposition presented in (26.23). It may be reduced to the following sum:

$$(26.25)\qquad \varphi(x_1, x_2, \ldots, x_n) = \sum_{a \in \mathbf{A}} \left[\prod_{j=1}^{n} x_j^{a_j}(1 - x_j)^{1-a_j} \right],$$

where the sum is extended to the set **A** of n-tuples (a) such that $\varphi(x) = 1$; this relation is identical, to within notation, to relation (22.18).

Example. Consider the case of a monotone structure function, but formulas (26.24) and (26.25) remain valid for monotone functions; let

$$\varphi(x_1, x_2, x_3) = x_1\, x_3 + x_2\, x_3 - x_1\, x_2\, x_3 \,. \tag{26.26}$$

In Fig. 26.10 we have given the values of φ; this function takes the value 1 for the following 3-tuples:

$$(0, 1, 1)\,, \quad (1, 0, 1)\,, \quad \text{and} \quad (1, 1, 1)\,. \tag{26.27}$$

x_1	x_2	x_3	φ
0	0	0	0
0	0	1	0
0	1	0	0
0	1	1	1
1	0	0	0
1	0	1	1
1	1	0	0
1	1	1	1

FIG. 26.10.

Using (26.25), we have

$$\begin{aligned} \varphi(x_1, x_2, x_3) &= (1 - x_1)\, x_2\, x_3 + x_1(1 - x_2)\, x_3 + x_1\, x_2\, x_3 \\ &= x_1\, x_3 + x_2\, x_3 - x_1\, x_2\, x_3 \,. \end{aligned} \tag{26.28}$$

Monotone Linear Composition

Theorem 26.V. *A necessary and sufficient condition for a structure function $\varphi(x_1, x_2, \ldots, x_n)$ to be monotone is that it be formed by a linear composition of two monotone structure functions $\varphi_1(x_1, x_2, \ldots, x_{n-1})$ and $\varphi_2(x_1, x_2, \ldots, x_{n-1})$ of order at most equal to $n - 1$:*

(26.29)

$$\varphi(x_1, x_2, \ldots, x_n) = x_n\, \varphi_1(x_1, x_2, \ldots, x_{n-1}) + (1 - x_n)\, \varphi_2(x_1, x_2, \ldots, x_{n-1})$$

where the functions φ_1 and φ_2 satisfy

$$\forall (x_1, x_2, \ldots, x_{n-1}) : \varphi_1(x_1, x_2, \ldots, x_{n-1}) \geqslant \varphi_2(x_1, x_2, \ldots, x_{n-1})\,. \tag{26.30}$$

PROOF. From (26.23) we have

(26.31)

$$\varphi(x_1, x_2, \ldots, x_n) \equiv x_n\, \varphi(x_1, x_2, \ldots, x_{n-1}, 1) + (1 - x_n)\, \varphi(x_1, x_2, \ldots, x_{n-1}, 0).$$

Put

$$\varphi_1(x_1, x_2, \ldots, x_{n-1}) = \varphi(x_1, x_2, \ldots, x_{n-1}, 1)\,, \tag{26.32}$$

$$\varphi_2(x_1, x_2, \ldots, x_{n-1}) = \varphi(x_1, x_2, \ldots, x_{n-1}, 0)\,, \tag{26.33}$$

which results in

$$\varphi(x_1, x_2, \ldots, x_{n-1}, x_n) = x_n[\varphi_1 - \varphi_2] + \varphi_2\,. \tag{26.34}$$

By successively using r-tuples such that

$$x_i^{(1)} = x_i^{(2)}\,, \qquad i = 1, \ldots, n-1\,; \qquad x_n^{(1)} = 1\,; \qquad x_n^{(2)} = 0\,;$$

then $\quad x_i^{(1)} \geqslant x_i^{(2)}\,, \quad i = 1, \ldots, n-1\,; \quad x_n^{(1)} = x_n^{(2)} = 0\,;$

and finally $\quad x_i^{(1)} \geqslant x_i^{(2)}\,, \quad i = 1, \ldots, n-1\,; \quad x_n^{(1)} = x_n^{(2)} = 1\,;$

we easily see that

$$\varphi \text{ is monotone} \quad \Leftrightarrow \quad \begin{cases} \varphi_1 \text{ and } \varphi_2 \text{ are monotone} \\ \varphi_1 \geqslant \varphi_2 \text{ for all } (x_1, x_2, \ldots, x_{n-1}). \end{cases} \tag{26.35}$$

Linear Composition of Networks. Let $\mathfrak{R}_1$ and $\mathfrak{R}_2$ be reliability networks equivalent, respectively, to φ_1 and φ_2. Then the network of Fig. 26.11 is equivalent to the structure function φ given by (26.29): If e_n is in a good state ($x_n = 1$), the system functions if and only if the substructure $\mathfrak{R}_1$ functions ($\varphi_1 = 1$). If e_n has failed, the system functions if and only if the two substructures $\mathfrak{R}_1$ and $\mathfrak{R}_2$ are in good states; but $\varphi_1 \geqslant \varphi_2$ implies that if $\mathfrak{R}_2$ functions, $\mathfrak{R}_1$ also functions; the only condition therefore is that $\varphi_2 = 1$.

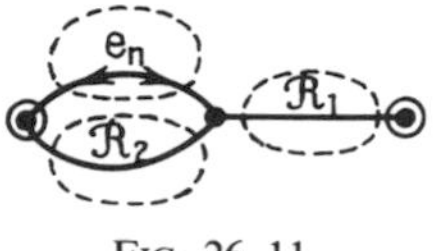

FIG. 26.11.

We may also show the equivalence by writing the structure function of the network $\mathfrak{R}$ by the method of links (Section 22, Eq. (22.1)):

$$\varphi_r = 1 - (1 - x_n\, \varphi_1)(1 - \varphi_1\, \varphi_2)\,,$$

from which

$$\varphi_r = x_n\, \varphi_1 + (1 - x_n)\, \varphi_1\, \varphi_2\,.$$

Since $\varphi_1 \geqslant \varphi_2$, we have $\varphi_1\varphi_2 = \varphi_2$, and (26.29) is recovered.

Remarks. (1) Monotone linear composition may be interpreted in the following fashion: As long as the component e_n functions, the system behaves as the substructure φ_1. The failure of e_n degrades the reliability of the system, which becomes that of the substructure φ_2 (with $\varphi_2 \leqslant \varphi_1$): if $\varphi_1 = 1$ and $\varphi_2 = 0$ at the moment of the failure of e_n, this will entail the failure of the system; if $\varphi_1 = 1$ and $\varphi_2 = 1$, then failure of φ_2 will entail that of the system.

(2) The substructures φ_1 and φ_2 in principle have the same components $(e_1, e_2, \ldots, e_{n-1})$, but certain of these may be useless in one or the other of the substructures φ_1 and φ_2 with the reservation that (26.30) be satisfied, that is, that any link of $\mathcal{R}_2$ is a link of $\mathcal{R}_1$, and any cut of $\mathcal{R}_1$ is a cut of $\mathcal{R}_2$. In the limit, $\mathcal{R}_1$, $\mathcal{R}_2$ (and e_n) may be modules of $\mathcal{R}$ (Examples 1 and 2 below).

Examples.

(1) If $\varphi_2 \equiv 0$ (degenerate structure of order 0), inequality (26.30) is satisfied for any φ_1, and $\varphi(x_1, \ldots, x_n) = x_n \varphi_1(x_1, \ldots, x_{n-1})$. The component e_n is in series with the module φ_1 (or $\mathcal{R}_1$; in Fig. 26.11, $\mathcal{R}_2$ has no links).

(2) Similarly, if $\varphi_1 \equiv 1$ (another degenerate structure of order 0), (26.30) is satisfied for any φ_2, and $\varphi(x_1, \ldots, x_n) = x_n + (1 - x_n)\varphi_2(x_1, \ldots, x_{n-1})$. The component e_n is in parallel with the module φ_2; in Fig. 26.11, $\mathcal{R}_1$ has no cuts.

(3) Consider again the example (Fig. 26.12) given in (26.26):

$$\varphi(x_1, x_2, x_3) = x_1 x_3 + x_2 x_3 - x_1 x_2 x_3 . \tag{26.36}$$

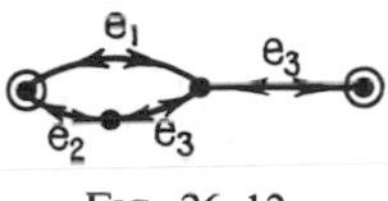

FIG. 26.12.

This structure function may be written as

$$\varphi(x_1, x_2, x_3) = x_1 x_3 + (1 - x_1) x_2 x_3 .$$

Put

$$\varphi_1(x_2, x_3) = x_3 , \qquad \varphi_2(x_2, x_3) = x_2 x_3 . \tag{26.37}$$

Then φ_1 is evidently monotone, and likewise φ_2; on the other hand, $\varphi_1 \geqslant \varphi_2$, as one may see in the table of values:

$$\begin{aligned} &\varphi_1(0, 0) = 0 , \qquad &\varphi_2(0, 0) = 0 , \\ &\varphi_1(0, 1) = 1 , \qquad &\varphi_2(0, 1) = 0 , \\ &\varphi_1(1, 0) = 0 , \qquad &\varphi_2(1, 0) = 0 , \\ &\varphi_1(1, 1) = 1 , \qquad &\varphi_2(1, 1) = 1 . \end{aligned} \tag{26.38}$$

The two conditions on the member on the right of (26.35) are indeed satisfied. One may confirm that (26.36) is a monotone structure function.

Determination of the Set of Monotone Functions through Recurrence. Theorem 26.V shows that linear composition allows us to determine, using recursion, the set of monotone structure functions of order at most equal to n.

By way of an example, we determine the monotone structures of orders 0, 1, 2, and 3. The monotone structures of order 0 are the two degenerate structures

$$\varphi_1^{(0)} = 1 \quad \text{and} \quad \varphi_2^{(0)} = 0. \tag{26.39}$$

Since $\varphi_1^{(0)} > \varphi_2^{(0)}$, we obtain the monotone structure of order 1:

$$\varphi_1^{(1)}(x_1) = x_1\, \varphi_1^{(0)} + (1 - x_1)\, \varphi_2^{(0)} = x_1 \,. \tag{26.40}$$

We now have at our disposal three monotone functions: $\varphi_1^{(0)}$, $\varphi_2^{(0)}$, and $\varphi_1^{(1)}$, for which we may write

$$\forall (x_1) : \quad \varphi_1^{(0)} \geqslant \varphi_1^{(1)} \geqslant \varphi_2^{(0)} \,. \tag{26.41}$$

From property (26.41), considering the two pairs $(\varphi_1^{(0)}, \varphi_1^{(1)})$ and $(\varphi_1^{(1)}, \varphi_2^{(0)})$ (the third has already been used to form (26.40)), we obtain

$$\begin{aligned} \varphi_1^{(2)}(x_1, x_2) &= x_2 . \varphi_1^{(0)} + (1 - x_2) . \varphi_1^{(1)} \\ &= x_2 + (1 - x_2)\, x_1 \\ &= x_1 + x_2 - x_1\, x_2 , \end{aligned} \tag{26.42}$$

$$\begin{aligned} \varphi_2^{(2)}(x_1, x_2) &= x_2 . \varphi_1^{(1)} + (1 - x_2) . \varphi_2^{(0)} \\ &= x_2\, x_1 + (1 - x_2) . 0 \\ &= x_1\, x_2 \,. \end{aligned} \tag{26.43}$$

We order the functions already obtained:

$$\varphi_1^{(0)} \geqslant \varphi_1^{(2)} \geqslant \varphi_1^{(1)} \geqslant \varphi_2^{(2)} \geqslant \varphi_2^{(0)} \,. \tag{26.44}$$

There are $\binom{5}{2} - \binom{3}{2} = 10 - 3 = 7$ pairs of structure functions not already taken into account, namely:

$$\begin{aligned} &(\varphi_1^{(0)}, \varphi_1^{(2)}) , \quad (\varphi_1^{(0)}, \varphi_2^{(2)}) , \quad (\varphi_1^{(2)}, \varphi_1^{(1)}) , \quad (\varphi_1^{(2)}, \varphi_2^{(2)}) , \quad (\varphi_1^{(2)}, \varphi_2^{(0)}) , \\ &(\varphi_1^{(1)}, \varphi_2^{(2)}) , \quad (\varphi_2^{(2)}, \varphi_2^{(0)}) , \end{aligned} \tag{26.45}$$

from which

$$(26.46) \quad \varphi_1^{(3)}(x_1, x_2, x_3) = x_3\,\varphi_1^{(0)} + (1 - x_3)\,\varphi_1^{(2)}$$
$$= x_3 + (1 - x_3)(x_1 + x_2 - x_1 x_2)$$
$$= x_1 + x_2 + x_3 - x_1 x_2 - x_1 x_3 - x_2 x_3 + x_1 x_2 x_3\,,$$

$$(26.47) \quad \varphi_2^{(3)}(x_1, x_2, x_3) = x_3\,\varphi_1^{(0)} + (1 - x_3)\,\varphi_2^{(2)}$$
$$= x_3 + (1 - x_3)\,x_1 x_2$$
$$= x_3 + x_1 x_2 - x_1 x_2 x_3\,,$$

$$(26.48) \quad \varphi_3^{(3)}(x_1, x_2, x_3) = x_3\,\varphi_1^{(2)} + (1 - x_3)\,\varphi_1^{(1)}$$
$$= x_3(x_1 + x_2 - x_1 x_2) + (1 - x_3)\,x_1$$
$$= x_1 + x_2 x_3 - x_1 x_2 x_3\,,$$

$$(26.49) \quad \varphi_4^{(3)}(x_1, x_2, x_3) = x_3\,\varphi_1^{(2)} + (1 - x_3)\,\varphi_2^{(2)}$$
$$= x_3(x_1 + x_2 - x_1 x_2) + (1 - x_3)\,x_1 x_2$$
$$= x_1 x_2 + x_1 x_3 + x_2 x_3 - 2\,x_1 x_2 x_3\,,$$

$$(26.50) \quad \varphi_5^{(3)}(x_1, x_2, x_3) = x_3\,\varphi_1^{(2)} + (1 - x_3)\,\varphi_2^{(0)}$$
$$= x_3(x_1 + x_2 - x_1 x_2) + (1 - x_3).0$$
$$= x_1 x_3 + x_2 x_3 - x_1 x_2 x_3,$$

$$(26.51) \quad \varphi_6^{(3)}(x_1, x_2, x_3) = x_3\,\varphi_1^{(1)} + (1 - x_3)\,\varphi_2^{(2)}$$
$$= x_3 x_1 + (1 - x_3)\,x_1 x_2$$
$$= x_1 x_2 + x_1 x_3 - x_1 x_2 x_3,$$

$$(26.52) \quad \varphi_7^{(3)}(x_1, x_2, x_3) = x_3\,\varphi_2^{(2)} + (1 - x_3).\varphi_2^{(0)}$$
$$= x_3 x_1 x_2 + (1 - x_3).0$$
$$= x_1 x_2 x_3\,.$$

Notice moreover that $\varphi_2^{(3)}$ and $\varphi_3^{(3)}$ may be obtained from one another by a circular permutation of x_1, x_2, x_3; similarly, $\varphi_5^{(3)}$ and $\varphi_6^{(3)}$. Finally, there remain five distinct types of structure.

The table presented in Fig. 26.13 gives the corresponding reliability networks for the results obtained above. The reader should compare these results with those obtained in Section 21, Figs. 21.6–21.8, by the method of enumeration with the aid of free distributive lattices on n generators.

Linear Composition of Reliability Functions. The reliability function $h(p)$, being identical to the simple form of the structure function $\varphi(x)$ (see the presentations of Sections 22 and 25), that is,

$$(26.53) \qquad h(p) \equiv \varphi_s(p) \quad \text{where } \varphi_s(x) \text{ is the simple form,}$$

Order of the structure	Structure function	Minimal links	Reliability network
0	$\varphi_1^{(0)} = 1$		A 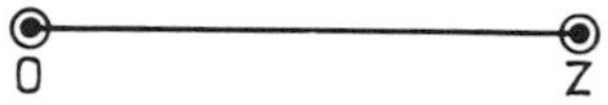 (O, Z)
	$\varphi_2^{(0)} = 0$		B (O, Z)
1	$\varphi_1^{(1)}(x_1) = x_1$	$\{e_1\}$	C 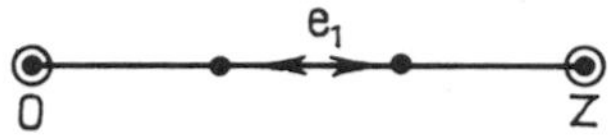 (O, e_1, Z)
2	$\varphi_1^{(2)}(x_1, x_2) = x_1 + x_2 - x_1 x_2$	$\{e_1\}$ $\{e_2\}$	D 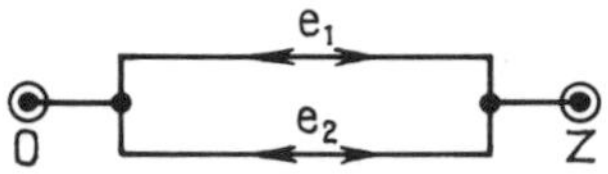 (O, e_1, e_2, Z)
	$\varphi_2^{(2)}(x_1, x_2) = x_1 x_2$	$\{e_1, e_2\}$	E  (O, e_1, e_2, Z)
3	$\varphi_1^{(3)}(x_1, x_2, x_3) = = x_1 + x_2 + x_3 - x_1 x_2 - x_1 x_3 - x_2 x_3 + x_1 x_2 x_3$	$\{e_1\}$ $\{e_2\}$ $\{e_3\}$	F 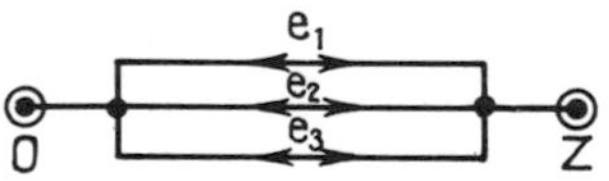 (O, e_1, e_2, e_3, Z)
	$\varphi_2^{(3)}(x_1, x_2, x_3) = x_3 + x_1 x_2 - x_1 x_2 x_3$	$\{e_3\}$ $\{e_1, e_2\}$	G 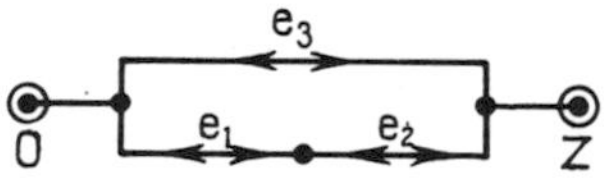 (O, e_3, e_1, e_2, Z)
	$\varphi_4^{(3)}(x_1, x_2, x_3) = x_1 x_2 + x_1 x_3 + x_2 x_3 - 2 x_1 x_2 x_3$	$\{e_1, e_2\}$ $\{e_1, e_3\}$ $\{e_2, e_3\}$	H 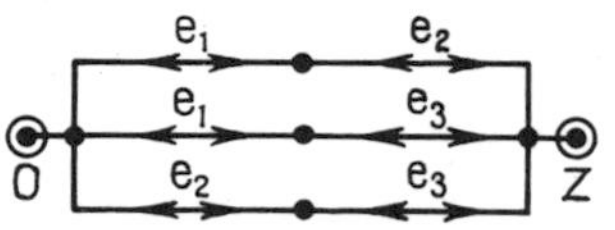 (O, e_1, e_2, e_1, e_3, e_2, e_3, Z)
	$\varphi_5^{(3)}(x_1, x_2, x_3) = x_1 x_3 + x_2 x_3 - x_1 x_2 x_3$	$\{e_1, e_3\}$ $\{e_2, e_3\}$	I 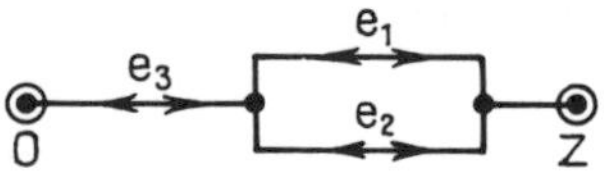 (O, e_3, e_1, e_2, Z)
	$\varphi_7^{(3)}(x_1, x_2, x_3) = x_1 x_2 x_3$	$\{e_1, e_2, e_3\}$	J 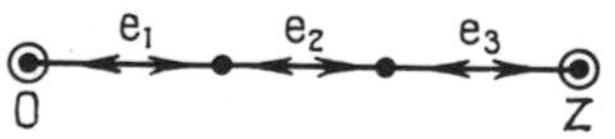 (O, e_1, e_2, e_3, Z)

FIG. 26.13.

the properties that arose in their study remain valid for reliability functions since the polynomials obtained by linear composition are of first degree with respect to each of the variables.

In particular, we may obtain by recurrence the set of reliability functions for monotone structures. In the particular case where

(26.54) $$p_1 = p_2 = \cdots = p_n = p\,,$$

we have

(26.55) Order 0 : $$h_1^{(0)}(p) = 1\,,$$

(26.56) $$h_2^{(0)}(p) = 0 \qquad (h_1^{(0)} > h_2^{(0)})\,.$$

(26.57) Order 1 : $$h_1^{(1)}(p) = ph_1^{(0)} + (1 - p)\, h_2^{(0)} = p \qquad (h_1^{(0)} \geqslant h_1^{(1)} \geqslant h_2^{(0)})\,.$$

(26.58) Order 2 : $$h_1^{(2)}(p) = ph_1^{(0)} + (1 - p)\, h_1^{(1)} = 2\,p - p^2\,,$$

(26.59) $$h_2^{(2)}(p) = ph_1^{(1)} + (1 - p)\, h_2^{(0)} = p^2\,.$$

(26.60) Order 3 : $$h_1^{(3)}(p) = 3\,p - 3\,p^2 + p^3\,,$$

(26.61) $$h_2^{(3)}(p) = h_3^{(3)}(p) = p + p^2 - p^3\,,$$

(26.62) $$h_4^{(3)}(p) = 3\,p^2 - 2\,p^3\,,$$

(26.63) $$h_5^{(3)}(p) = h_6^{(3)}(p) = 2\,p^2 - p^3\,,$$

(26.64) $$h_7^{(3)}(p) = p^3\,.$$

27 Representative Curves of Reliability Functions for Monotone Structures. Theorem of Moore and Shannon

We suppose in the remainder of this section that the components e_1, $e_2, \ldots, e_n$ have the same reliability: $p_1 = p_2 = \cdots = p_n = p$.

Theorem 27.I (Theorem of Moore and Shannon). *The reliability function of a monotone structure of order n all of whose components have the same reliability p satisfies the following inequalities:*

(27.1) $$\frac{\mathrm{d}}{\mathrm{d}p}\,h(p) \leqslant \sqrt{\frac{nh(p)\,[1 - h(p)]}{p(1 - p)}}\,,$$

(27.2) $$\frac{\mathrm{d}}{\mathrm{d}p}\,h(p) \geqslant \frac{h(p)\,[1 - h(p)]}{p(1 - p)}\,.$$

The relations result directly from (25.32) and (25.33), taking into account the relation

(27.3) $$\frac{\mathrm{d}}{\mathrm{d}p}h(p) = \sum_{i=1}^{n} \frac{\partial h(p_1, \ldots, p_n)}{\partial p_i}\,.$$

These inequalities were obtained by Moore and Shannon[5] in 1956. We review briefly the proof that they have for inequality (27.2), now called the "Moore–Shannon theorem."

We shall reason by recurrence. For any reliability function $h(p)$ of a monotone structure of order n, one may write (cf. Section 26, linear composition)

$$h(p) = pf(p) + (1 - p)\, g(p) \quad \text{with}^{6} \quad f(p) \geqslant g(p)\,, \tag{27.4}$$

or more briefly

$$h = pf + (1 - p)\, g\,. \tag{27.5}$$

Using the evident inequality

$$(1 - p).p.(f - g)(1 - f + g) \geqslant 0\,, \qquad 0 \leqslant p \leqslant 1\,, \tag{27.6}$$

we may write, after several transformations,

$$(1-f)\,pf + (1-p)\,pf + (1-p)\,g(1-g) - (1-p)\,pg \geqslant h(1-h)\,. \tag{27.7}$$

If f and g have the property

$$\frac{\mathrm{d}f}{\mathrm{d}p} \geqslant \frac{f(1 - f)}{p(1 - p)}\,, \tag{27.8}$$

we obtain

$$p\,\frac{\mathrm{d}f}{\mathrm{d}p} + f + (1 - p)\,\frac{\mathrm{d}g}{\mathrm{d}p} - g \geqslant \frac{h(1 - h)}{p(1 - p)}\,, \tag{27.9}$$

and therefore

$$\frac{\mathrm{d}h}{\mathrm{d}p} = p\,\frac{\mathrm{d}f}{\mathrm{d}p} + f + (1 - p)\,\frac{\mathrm{d}g}{\mathrm{d}p} - g \geqslant \frac{h(1 - h)}{p(1 - p)}\,. \tag{27.10}$$

However $h_1^{(0)}(p) = 1$, $h_2^{(0)}(p) = 0$, $h_1^{(1)}(p) = p$ possess property (27.8); the theorem is therefore proved by recurrence.

Representative Curves for Reliability Functions of Monotone Structures with Components of Equal Reliabilities. The Moore–Shannon theorem (27.2) shows that the derivative of $h(p)$ is bounded below by solutions of the differential equation

$$p(1 - p)\,\frac{\mathrm{d}u(p)}{\mathrm{d}p} - u(p)\,[1 - u(p)] = 0\,. \tag{27.11}$$

[5] In an article published in 1956, Moore and Shannon [40] studied redundance for relays (cf. Chapter VI). This theorem was thus discovered within the framework of the solution of a concrete problem.

[6] This condition applies according to relation (26.30) of Theorem 26.V.

The family of solutions to this equation is

$$u_c(p) = \frac{cp}{1 + (c - 1)p} \tag{27.12}$$

where c is an arbitrary constant. In fact, we shall consider only those values of $c \geqslant 0$ for which $u_c(p) \geqslant 0$ for $0 \leqslant p \leqslant 1$. For any $c \neq 0$, we have $u_c(0) = 0$ and $u_c(1) = 1$; Fig. 27.1 represents the shape of the curves $u_c(p)$, which are branches of hyperbolas.

The three reliability functions of order 0 or 1 coinciding, respectively, with $u_c(p)$ for $c = +\infty$, $c = 0$, and $c = 1$ are

$$h_1^{(0)}(p) = 1, \qquad u_\infty(p) = 1, \tag{27.13}$$

$$h_2^{(0)}(p) = 0, \qquad u_0(p) = 0, \tag{27.14}$$

$$h_1^{(1)}(p) = p, \qquad u_1(p) = p. \tag{27.15}$$

Now consider a reliability function of order at least 2. For a given value p_1 of p, we have

$$h(p_1) = u_{c_1}(p_1) \tag{27.16}$$

for a value c_1 given by

$$c_1 = \frac{1 - p_1}{p_1} \frac{h(p_1)}{1 - h(p_1)}. \tag{27.17}$$

At this point $(p_1, h(p_1))$, the Moore–Shannon theorem shows that $\mathrm{d}h/\mathrm{d}p \geqslant \mathrm{d}u_{c_1}/\mathrm{d}p$: the curve $h(p)$ traverses the curve u_{c_1} from below. Moreover, we have $h(p) > u_{c_1}(p)$ for $p > p_1$; that is, $h(p)$ does not cut again $u_{c_1}(p)$. If it did we would have either $h(p) = u_{c_1}(p)$ in a nonempty interval, which is impossible since $h(p)$ is a polynomial of degree at least 2 and $u_{c_1}(p)$ is a hyperbola or a straight line, or else $\mathrm{d}h/\mathrm{d}p < \mathrm{d}u_{c_1}/\mathrm{d}p$ at a new point of contact. As a consequence, a reliability function of order at least equal to 2 meets at most

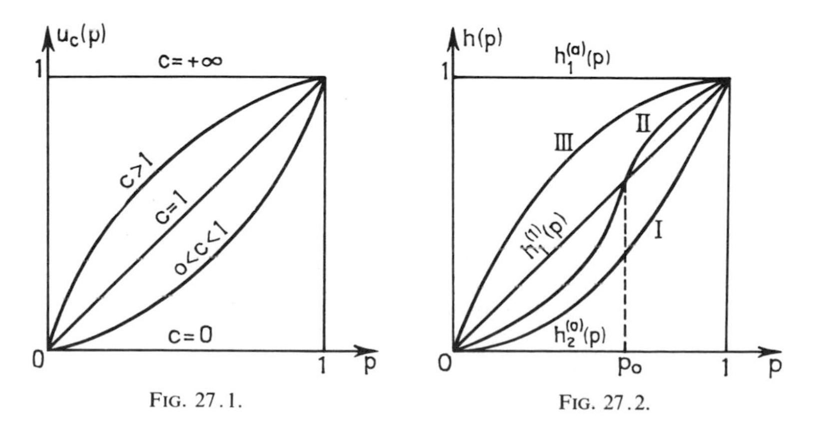

FIG. 27.1. FIG. 27.2.

one time any given $u_c(p)$ curve. This is applicable in particular to $u_1(p)$, that is, to the first bisector.

To stress this property we classify the reliability functions of monotone structures with components of equal reliability into three types (excluding functions of order less than 2):

$$\text{TYPE I}: \quad h(p) < p \quad \text{for} \quad 0 < p < 1 .$$

$$\text{TYPE II}: \quad \begin{aligned} h(p) &< p && \text{for} \quad 0 < p < p_0, \\ &= p_0 && \text{for} \quad p = p_0, \\ &> p && \text{for} \quad p_0 < p < 1 . \end{aligned}$$

$$\text{TYPE III}: \quad h(p) > p \quad \text{for} \quad 0 < p < 1 .$$

These three types of curves are presented in Fig. 27.2; type II is called an S curve.

Identification of the Type of a Reliability Curve. To determine the type of a function, we find the slope at the origin and the slope at point $p = 1$. Let $h'(0)$ be the slope at the origin and $h'(1)$ that at point $p = 1$. The three possible cases are:

$$(27.18) \quad 0 \leqslant h'(0) < 1 , \quad 1 \leqslant h'(1), \quad \text{type I} ,$$

$$(27.19) \quad 0 \leqslant h'(0) < 1 , \quad 0 \leqslant h'(1) < 1 , \quad \text{type II} ,$$

$$(27.20) \quad 1 \leqslant h'(0) , \quad 0 \leqslant h'(1) < 1 , \quad \text{type III} .$$

Examples.

(1) Let

$$(27.21) \quad h(p) = p^3 \; ; \quad h'(p) = 3\,p^2 , \quad h'(0) = 0 , \quad h'(1) = 3 .$$

According to (27.18), the function is of type I.

(2) Let

$$(27.22) \quad h(p) = 3\,p^2 - 2\,p^3 \; ; \quad h'(p) = 6\,p - 6\,p^2 , \quad h'(0) = 0 , \quad h'(1) = 0 .$$

From (27.19), the function is of type II.

(3) Let

$$(27.23) \quad h(p) = 2\,p - p^2 , \quad h'(p) = 2 - 2\,p , \quad h'(0) = 2 , \quad h'(1) = 0 .$$

From (27.20), the function is of type III.

We may also obtain important information from the links and cuts of the structure function $\varphi(x_1, x_2, \ldots, x_n)$ corresponding to $h(p)$. Let $\varphi(x_1, x_2, \ldots, x_n)$ be the structure function of a system with n independent components; the corresponding reliability function is

$$(27.24) \quad h(p_1, p_2, \ldots, p_n) = E[\varphi(X_1, X_2, \ldots, X_n)] .$$

The recurrence relation

(27.25)

$$\varphi(x_1, x_2, \ldots, x_n) = x_n \cdot \varphi(x_1, x_2, \ldots, x_{n-1}, 1) + (1 - x_n)\, \varphi(x_1, x_2, \ldots, x_{n-1}, 0)$$

follows by taking the mathematical expectation:

(27.26)
$$h(p_1, p_2, \ldots, p_n) = p_n \cdot E[\varphi(X_1, X_2, \ldots, X_{n-1}, 1)] + (1 - p_n) \cdot E[\varphi(X_1, X_2, \ldots, X_{n-1}, 0)]\,.$$

Differentiating with respect to p_n, there follows

(27.27)
$$\frac{\partial}{\partial p_n} h(p_1, p_2, \ldots, p_n) = E[\varphi(X_1, X_2, \ldots, X_{n-1}, 1)] - E[\varphi(X_1, X_2, \ldots, X_{n-1}, 0)]\,.$$

This relation is valid for all components e_i, $i = 1, 2, \ldots, n$, and if $p_1 = p_2 = \cdots = p_n = p$, we obtain

(27.28)

$$\frac{\mathrm{d}}{\mathrm{d}p} h(p) = \sum_{i=1}^{n} \frac{\partial h(p_1, p_2, \ldots, p_n)}{\partial p_i} \cdot \frac{\partial p_i}{\partial p} = \sum_{i=1}^{n} \frac{\partial h(p_1, p_2, \ldots, p_n)}{\partial p_i}$$

$$= \sum_{i=1}^{n} \{ E[\varphi(X_1, \ldots, X_i = 1, \ldots, X_n)] - E[\varphi(X_1, \ldots, X_i = 0, \ldots, X_n)] \}\,.$$

Putting $p = 0$, relation (27.28) becomes

(27.29)
$$h'(0) = \sum_{i=1}^{n} [\varphi(0, \ldots, x_i = 1, \ldots, 0) - \varphi(0, \ldots, x_i = 0, \ldots, 0)]\,.$$

The derivative at the origin is thus equal to the number of links having a single component (this result and the next may also be deduced from the simple form of the reliability function—see (25.18) or (26.25)).

Likewise, if we put $p = 1$ in relation (27.28), we see that the derivative at the point $p = 1$ is equal to the number of cuts having a single component.

Finally, we obtain Table 27.1.

Number of links having a single component	Number of cuts having a single component	Type of reliability function
0	0	II (S curve)
$\geqslant 1$	0	III
0	$\geqslant 1$	I
1	1	bisector : $h(p) = p$

Note that we may not have at the same time a cut having a single component and a link having just one component; the slope $h(p)$ is thus zero for at least one of the two values $p = 0$ and $p = 1$ (with the exception of the function $h(p) = p$).

Examples.

(1) A series structure gives a reliability function of type I.

(2) A parallel structure gives a reliability function of type III.

(3) A bridge for which the edges correspond to different components gives a reliability function of type II.

28 Systems Monotone in Probability

Definition. *Let S be a system of order n for which the random structure function is $\varphi(X)$. We shall say that $\varphi(X)$ is monotone in probability if*

$$(28.1)\quad \mathrm{pr}\,\{\,\varphi(X) = 1 \mid S(X) = k\,\} \leqslant \mathrm{pr}\,\{\,\varphi(X) = 1 \mid S(X) = k + 1\,\}\,, \qquad k = 0, 1, 2, \ldots, n - 1\,,$$

where

$$(28.2)\qquad S(X) = X_1 + X_2 + \cdots + X_n\,,$$

the indices 1, 2, ..., *n representing the active components of the system, and where*

$$(28.3)\qquad \mathrm{pr}\,\{\,\varphi(X) = 1 \mid S(X) = k\,\} = \frac{\mathrm{pr}\,\{\,\varphi(X) = 1,\, S(X) = k\,\}}{\mathrm{pr}\,\{\,S(X) = k\,\}}\,,$$

following the usual definition of conditional probability.

Remarks.

(1) The above property involves both the structure function $\varphi(x)$ and the reliabilities of the components through the random function of structure (25.5).

(2) Useless components do not figure in the count of the components in good state.

(3) The reliability of a structure monotone in probability is increasing in the mean with the number of components in good state; it is, however, not excluded, in particular cases, that the return to functioning of a component could degrade the functioning of the system, that is, that the structure function of the system might be nonmonotone.

Example. Consider the structure function

$$(28.4)\qquad \varphi(x) = x_1 + x_2\, x_3 - x_1\, x_3\,.$$

The table of values of this function (central column of Fig. 28.1) shows that φ is not monotone; we have $\varphi(1, 0, 1) = 0$, but $\varphi(1, 0, 0) = 1$.

The conditional probabilities occurring in (28.1) are easily calculated

x_1	x_2	x_3	φ	$\text{pr}\,(X_1 = x_1, X_2 = x_2, X_3 = x_3)$
0	0	0	0	$(1 - p_1)(1 - p_2)(1 - p_3)$
0	0	1	0	$(1 - p_1)(1 - p_2)\,p_3$
0	1	0	0	$(1 - p_1)\,p_2\,(1 - p_3)$
1	0	0	1	$p_1\,(1 - p_2)(1 - p_3)$
0	1	1	1	$(1 - p_1)\,p_2\,p_3$
1	0	1	0	$p_1\,(1 - p_2)\,p_3$
1	1	0	1	$p_1\,p_2\,(1 - p_3)$
1	1	1	1	$p_1\,p_2\,p_3$

FIG. 28.1.

using (28.3) on one hand, and on the other the probabilities of state of the set of components, which appear in the right-hand part of Fig. 28.1. In conformity with the notation of Section 25, we have called the probabilities of good functioning of the three components p_1, p_2, and p_3.

Setting

$$h_1(p) = \text{pr}\,\{\,\varphi(X) = 1 \mid S(X) = 1\,\} \tag{28.5}$$

and

$$h_2(p) = \text{pr}\,\{\,\varphi(X) = 1 \mid S(X) = 2\,\}\,, \tag{28.6}$$

we obtain

$$h_1(p) = \frac{p_1(1 - p_2)(1 - p_3)}{(1 - p_1)(1 - p_2)\,p_3 + (1 - p_1)\,p_2(1 - p_3) + p_1(1 - p_2)(1 - p_3)} \tag{28.7}$$

$$h_2(p) = \frac{(1 - p_1)\,p_2\,p_3 + p_1\,p_2(1 - p_3)}{(1 - p_1)\,p_2\,p_3 + p_1(1 - p_2)\,p_3 + p_1\,p_2(1 - p_3)}\,. \tag{28.8}$$

The system is monotone in probability if and only if

$$h_1(p) \leqslant h_2(p)\,, \tag{28.9}$$

which represents an equation of degree 6 with respect to the set of parameters p_1, p_2, and p_3. We shall content ourselves, by studying certain particular cases, with showing that the system may or may not be monotone in probability, according to the value of the vector $p = (p_1, p_2, p_3)$.

Suppose first that $p_1 = p_2 = p_3 = \alpha$, with $0 \leqslant \alpha \leqslant 1$. Then we have

$$h_1(\alpha, \alpha, \alpha) = \frac{\alpha(1 - \alpha)^2}{3\,\alpha(1 - \alpha)^2} = \frac{1}{3}\,, \tag{28.10}$$

$$h_2(\alpha, \alpha, \alpha) = \frac{2\,\alpha^2(1 - \alpha)}{3\,\alpha^2(1 - \alpha)} = \frac{2}{3}\,. \tag{28.11}$$

We see that for any such α, one has $h_1 < h_2$. In the space (p_1, p_2, p_3) where the set of possible values is the unit cube (Fig. 28.2), the system is monotone in probability on the principal diagonal joining the origin (0, 0, 0) to the vertex (1, 1, 1).

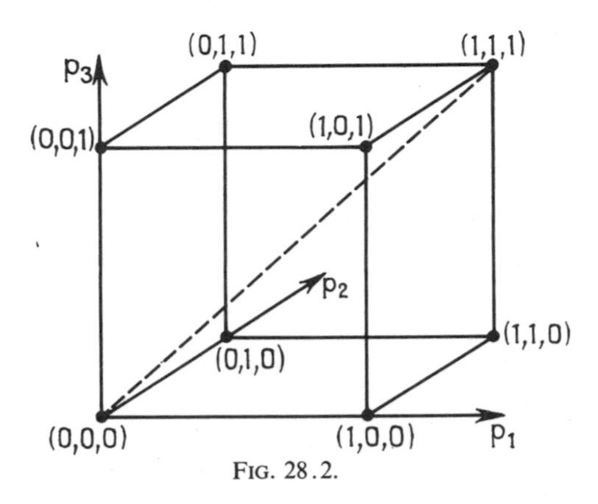

FIG. 28.2.

We now go on to see what happens on the faces of the cube, but it is necessary first to point out that $h_1(p)$ and $h_2(p)$ are not defined everywhere. For example, at the origin vertex ($p_1 = p_2 = p_3 = 0$), we have

$$\text{pr}\{S(X) = 0\} = 1, \quad \text{and} \quad \text{pr}\{S(X) = 1\} = \text{pr}\{S(X) = 2\} = 0;$$

it follows that the conditional probabilities (28.7) and (28.8) are not defined at this point (it is the same at the point $p_1 = p_2 = p_3 = 1$). Rigorously then the result that we obtained above is valid only on those points of the principal diagonal *interior* to the cube. The same difficulty is found on the edges of the cube, but $h_1(p)$ and $h_2(p)$ are defined on the faces of the cube, at least at those points interior to these faces; in fact, a face corresponds to a case where one of the components has probability 0 or 1 of functioning. The number of components able to function may then take either the values 0, 1, or 2 or the values 1, 2, or 3; in all cases we have $\text{pr}\{S(X) = 1\} > 0$ and $\text{pr}\{S(X) = 2\} > 0$ for points interior to the face. For example, if we set $p_2 = 0$ in (28.7) and (28.8) there results

$$(28.12) \qquad h_1(p_1, 0, p_3) = \frac{p_1(1 - p_3)}{(1 - p_1)p_3 + p_1(1 - p_3)},$$

$$(28.13) \qquad h_2(p_1, 0, p_3) = 0 \quad \text{for} \quad p_1 \neq 0 \quad \text{and} \quad p_3 \neq 0\,.$$

One sees that on this face the system is not monotone in probability. For reasons of continuity it is then necessarily the same at certain points interior to the cube.

Properties of Systems Monotone in Probability

Theorem 28.I. *Let S_1 and S_2 be two systems of order $n-1$ having the same components $e_1, e_2, \ldots, e_{n-1}$ with respective reliabilities $p_1, \ldots, p_{n-1}$. If these systems are monotone in probability, the system S having for structure function*

$$(28.14) \quad \varphi(x_1, \ldots, x_n) = x_n\, \varphi_1(x_1, \ldots, x_{n-1}) + (1 - x_n)\, \varphi_2(x_1, \ldots, x_{n-1}),$$

where φ_1 and φ_2 are the structure functions of S_1 and S_2, is monotone in probability whatever the reliability p_n of the supplementary component e_n. (Relation (28.14) corresponds to what in Section 26 we referred to as "linear composition of structures.")

Theorem 28.I is obvious if the system S is of order $n-1$, which occurs if the structure functions φ_1 and φ_2 are equivalent; the three systems S_1, S_2, and S are then in fact identical. We therefore suppose that S is of order n. Putting

$$(28.15) \quad X = (X_1, X_2, \ldots, X_n),$$

$$(28.16) \quad Y = (X_1, X_2, \ldots, X_{n-1}),$$

then, according to (28.14), we have

$$(28.17) \quad k=0, \quad \text{pr}\,\{\, \varphi(X)=1 \mid S(X)=0 \,\} = \text{pr}\,\{\, \varphi_2(Y)=1 \mid S(Y)=0 \,\},$$

$$(28.18) \quad k=1, \ldots, n-1,$$
$$\text{pr}\,\{\, \varphi(X)=1 \mid S(X)=k \,\} = p_n \cdot \text{pr}\,\{\varphi(X)=1 \mid S(X)=k,\, X_n=1 \,\}$$
$$+ (1-p_n) \cdot \text{pr}\,\{\, \varphi(X)=1 \mid S(X)=k,\, X_n=0 \,\}.$$

Relation (28.18) may also be written as

$$(28.19) \quad \text{pr}\,\{\, \varphi(X)=1 \mid S(X)=k \,\} = p_n \cdot \text{pr}\,\{\, \varphi_1(Y)=1 \mid S(Y)=k-1 \,\}$$
$$+ (1-p_n) \cdot \text{pr}\,\{\, \varphi_2(Y)=1 \mid S(Y)=k \,\}.$$

Condition (28.1) then may be written as:

(a) for $k=0$,

$$(28.20) \quad \text{pr}\,\{\, \varphi_2(Y) = 1 \mid S(Y) = 0 \,\} \leqslant p_n \cdot \text{pr}\,\{\, \varphi_1(Y) = 1 \mid S(Y) = 0 \,\}$$
$$+ (1 - p_n) \cdot \text{pr}\,\{\, \varphi_2(Y) = 1 \mid S(Y) = 1 \,\}$$

or

(28.21)

$$(1-p_n) \cdot \text{pr}\,\{\, \varphi_2(Y)=1 \mid S(Y)=0 \,\} \leqslant (1-p_n) \cdot \text{pr}\,\{\, \varphi_2(Y)=1 \mid S(Y)=1 \,\};$$

(b) for $1 \leqslant k \leqslant n-1$,

(28.22)

$$p_n.\mathrm{pr}\{\varphi_1(Y)=1 \mid S(Y)=k-1\}+(1-p_n).\mathrm{pr}\{\varphi_2(Y)=1 \mid S(Y)=k\},$$
$$\leqslant p_n.\mathrm{pr}\{\varphi_1(Y)=1 \mid S(Y)=k\}+(1-p_n).\mathrm{pr}\{\varphi_2(Y)=1 \mid S(Y)=k+1\}.$$

Relations (28.21) and (28.22) follow directly from the property of monotonicity in probability of S_1 and S_2.

Theorem 28.II. *Any system having a monotone structure function is monotone in probability whatever the reliability of its components.*

The theorem is proved by recurrence: The systems of order 0 are systems that have one of the two degenerate structure functions:

(28.23) $$\varphi_1^{(0)} = 1,$$

(28.24) $$\varphi_2^{(0)} = 0.$$

These are considered monotone in probability by definition. The only monotone structure of order 1 is (see Section 2)

(28.25) $$\varphi_1^{(1)} = x_1.$$

A system having this structure function is monotone in probability, whatever the reliability p_1, since

(28.26) $$\mathrm{pr}\{\varphi_1^{(1)}(X) = 1 \mid S(X) = 0\} = 0$$

and

(28.27) $$\mathrm{pr}\{\varphi_1^{(1)}(X) = 1 \mid S(X) = 1\} = 1.$$

We have seen in Section 26 that any monotone structure of order n may be put in the form (28.14) where φ_1 and φ_2 are monotone structure functions with order at most equal to $n-1$; Theorem 28.I then proves Theorem 28.II by recurrence.

Theorem 28.III. *A system of order n all of whose components have the same reliability p is monotone in probability if and only if*

(28.28) $$(n-k)\,A_k \leqslant (k+1)\,A_{k+1}, \qquad k = 0, 1, \ldots, n-1,$$

where A_k is the number of links having k components.

Indeed, for a system all of whose components have the same reliability, one has, using the same reasoning as in obtaining (25.18),

(28.29) $$\mathrm{pr}\{S(X) = k\} = \binom{n}{k} p^k (1-p)^{n-k}$$

and

$$\text{pr}\{\varphi(X) = 1, S(X) = k\} = A_k p^k(1 - p)^{n-k}, \tag{28.30}$$

from which, according to (28.3),

$$\text{pr}\{\varphi(X) = 1 \mid S(X) = k\} = A_k / \binom{n}{k}. \tag{28.31}$$

Condition (28.1) ensuring that a system be monotone in probability may therefore be written

$$\frac{A_k}{\binom{n}{k}} \leqslant \frac{A_{k+1}}{\binom{n}{k+1}}, \qquad k = 0, 1, \ldots, n - 1. \tag{28.32}$$

By replacing $\binom{n}{k}$ with its valuation

$$\binom{n}{k} = \frac{n!}{k!\,(n-k)!}, \tag{28.33}$$

one obtains (28.28).

Note that by combining Theorems 28.III and 21.VI one may recover Theorem 28.II in the particular case of components with equal reliabilities.

Theorem 28.IV. *A system that is monotone in probability and whose components all have the same reliability p has a reliability function $h(p)$ such that*

$$\frac{\mathrm{d}h(p)}{\mathrm{d}p} \geqslant 0. \tag{28.34}$$

Indeed, the reliability function may be expressed (see (25.20))

$$h(p) = \sum_{k=0}^{n} A_k p^k(1 - p)^{n-k} \tag{28.35}$$

and its derivative may be written in the form

$$\frac{\mathrm{d}h(p)}{\mathrm{d}p} = \sum_{k=0}^{n} [(k + 1) A_{k+1} - (n - k) A_k]\, p^k(1 - p)^{n-k}. \tag{28.36}$$

This expression is nonnegative according to (28.28).

One may compare this theorem with Theorem 25.I, which thus appears valid for systems with nonmonotone structure functions, if there exist some values of p for which the system is monotone in probability, and only for these values of p. Recall that, for monotone structures, Theorem 27.I gives a better inferior limit.

29 Survival Function of a System

As we announced in Section 24, we now return to time considerations which were the essence of our discussions in Chapters 1 and 2, but here we shall be concerned with complex systems.

The reliability function $h(p)$ defined in Section 25 represents the probability that the system will be in a good state at the instant t as a function of the reliabilities p_i of its components, but the reliability p_i of the component e_i in the interval $(0, t)$ is nothing but the value $v_i(t)$ of its survival function at the instant t. The probability of functioning for the system at the instant t is therefore

$$h[v_1(t), v_2(t), \dots, v_n(t)] .$$

We have already remarked, however, in Section 24 that, in the case of a system with nonmonotone structure function, this quantity is not equal to the survival function $v(t)$ of the system. In order to obtain the latter it is necessary to calculate the probability that the system fails *for the first time* at the instant t; now this is the probability that is interesting since one generally requires from a system that it be capable of functioning without interruption during the interval $(0, t)$. Moreover, we remark that although the expression given by the reliability function for the probability of functioning at the instant t is theoretically valid for a nonmonotone system, it is debatable whether it is so in practice for such systems; indeed, this would assume implicitly that after a first failure of the system, the components still in a good state continue to age in the same fashion since their survival function $v_i(t)$ is given a priori. This is a particular expression of the general hypothesis of independence of the lifetimes of components (Section 24), which we have already indicated as very constraining, but whose validity must be viewed still more cautiously whenever the system is no longer in a functioning state.

We now go on to see through an example that the determination of the survival function of a nonmonotone system is complicated; we do not know a general method that allows one to obtain it simply. We shall then pass to the fortunately more manageable case of monotone structures.

Example (nonmonotone structure). Consider the structure function

$$\varphi(x) = x_2 - x_1 x_2 + x_1 x_2 x_3 . \tag{29.1}$$

Figure 29.1 represents the lattice of states of the set of components (cf. Section 15, p. 59); the vertices for which $\varphi = 1$ are represented by a dot, and those for which $\varphi = 0$ by a cross. The structure is not monotone since $\varphi(1, 1, 0) = 0$, whereas $\varphi(0, 1, 0) = 1$.

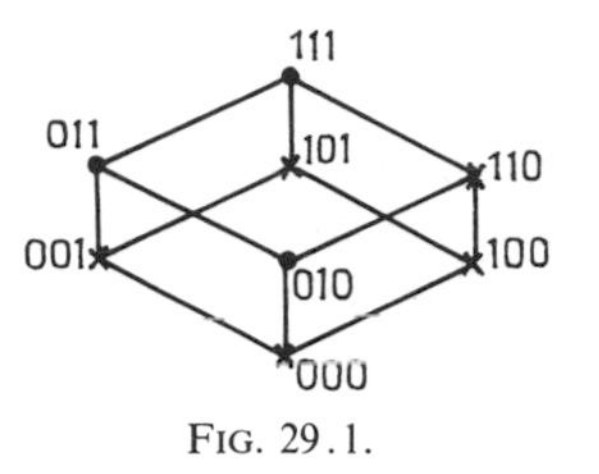

FIG. 29.1.

We examine the various ways that the system may fail for the first time. For example, if the component e_1 is the first to fail, the system continues to function (vertex 011); if e_2 then fails, the system no longer functions (vertex 001). With each path going from vertex (111) to vertex (000) in Fig. 29.1, that is, with each permutation of the three components defining an order in which they fail, one may associate the index of the component that brings about the failure of the system. Designating by T_1, T_2, and T_3 the lifetimes of the components, and by T the lifetime of the system, we may thus obtain the following implications:

(29.2)
$$\begin{aligned} T_1 < T_2 < T_3 &\Rightarrow T = T_2 \\ T_1 < T_3 < T_2 &\Rightarrow T = T_2 \\ T_2 < T_1 < T_3 &\Rightarrow T = T_2 \\ T_2 < T_3 < T_1 &\Rightarrow T = T_2 \\ T_3 < T_1 < T_2 &\Rightarrow T = T_3 \\ T_3 < T_2 < T_1 &\Rightarrow T = T_3 . \end{aligned}$$

Let E be the event

(29.3)
$$E = \{ T_3 = \min (T_1, T_2, T_3) \}$$

and $\bar{E}$ the complementary event. We see that, if E occurs, we have $T = T_3$; if $\bar{E}$ is the case, then $T = T_2$. The probability density $i(t)$ of T is therefore given by

(29.4)
$$i(t)\,\mathrm{d}t = \mathrm{pr}\,\{E\}.\mathrm{pr}\,\{t \leqslant T_3 < t+\mathrm{d}t \mid E\} + \mathrm{pr}\,\{\bar{E}\}.\mathrm{pr}\,\{t \leqslant T_2 \leqslant t+\mathrm{d}t \mid \bar{E}\},$$

or

(29.5)
$$\begin{aligned} i(t)\,\mathrm{d}t &= \mathrm{pr}\,\{ t \leqslant T_3 \leqslant t+\mathrm{d}t, E \} + \mathrm{pr}\,\{ t \leqslant T_2 \leqslant t+\mathrm{d}t, \bar{E} \} \\ &= \mathrm{pr}\,\{ t \leqslant T_3 \leqslant t+\mathrm{d}t \}.\mathrm{pr}\,\{ E \mid T_3 = t \} \\ &\quad + \mathrm{pr}\,\{ t \leqslant T_2 \leqslant t+\mathrm{d}t \}.\mathrm{pr}\,\{ \bar{E} \mid T_2 = t \} . \end{aligned}$$

Writing $i_k(t)$ for the probability density and $v_k(t)$ for the survival function of component k, we thus obtain

(29.6)
$$i(t) = i_3(t).v_1(t).v_2(t) + i_2(t)\left[1 - \int_0^t i_3(u).v_1(u)\,\mathrm{d}u\right].$$

The survival function $v(t)$ of the system may be obtained by integrating $i(t)$ between t and infinity.

Suppose, in order to carry out the calculations somewhat further, that the components each follow an exponential law (Section 6, p. 16):

$$v_k(t) = e^{-\lambda_k t}, \tag{29.7}$$

$$i_k(t) = \lambda_k e^{-\lambda_k t}. \tag{29.8}$$

After some calculation, one then finds

$$i(t) = \frac{\lambda_1}{\lambda_1 + \lambda_3} \lambda_2 e^{-\lambda_2 t} + \frac{\lambda_3}{\lambda_1 + \lambda_3} (\lambda_1 + \lambda_2 + \lambda_3) e^{-(\lambda_1 + \lambda_2 + \lambda_3) t}, \tag{29.9}$$

from which

$$v(t) = \frac{\lambda_1}{\lambda_1 + \lambda_3} e^{-\lambda_2 t} + \frac{\lambda_3}{\lambda_1 + \lambda_3} e^{-(\lambda_1 + \lambda_2 + \lambda_3) t}. \tag{29.10}$$

This probability law is called the "hyperexponential law"; it may be interpreted in the following fashion: There is a probability $\lambda_1/(\lambda_1 + \lambda_3)$ that the lifetime T of the system follows an exponential law with parameter λ_2 (law of T_2), and the complementary probability $\lambda_3/(\lambda_1 + \lambda_3)$ that T follows an exponential law with parameter $\lambda_1 + \lambda_2 + \lambda_3$; this second law is that of the random variable $\min(T_1, T_2, T_3)$. This interpretation suggests another way to calculate the survival function of the system.

Returning to the case where the components follow arbitrary laws, one may indeed write, by regrouping the first three cases of (29.2) and then the last three,

$$\begin{aligned} v(t) = {} & \mathrm{pr}\{T_3 > T_1\} \cdot \mathrm{pr}\{T_2 > t \mid T_3 > T_1\} \\ & + \mathrm{pr}\{T_3 < T_1\} \cdot \mathrm{pr}\{\min(T_1, T_2, T_3) > t \mid T_3 < T_1\}. \end{aligned} \tag{29.11}$$

Designate by A and B the two terms to the right-hand side of the equation. The first term A may be calculated easily since T_2 is independent in probability of T_1 and T_3:

$$A = \mathrm{pr}\{T_3 > T_1\} \cdot \mathrm{pr}\{T_2 > t\} = v_2(t) \int_0^\infty i_1(u)\, v_3(u)\, du. \tag{29.12}$$

The second term is a little more complicated to calculate. We first rewrite it in the form

(29.13)

$$\begin{aligned} B &= \mathrm{pr}\{T_3 < T_1, \min(T_1, T_2, T_3) > t\} \\ &= \mathrm{pr}\{\min(T_1, T_2, T_3) > t\} \cdot \mathrm{pr}\{T_3 < T_1 \mid \min(T_1, T_2, T_3) > t\}. \end{aligned}$$

The event that figures in the first factor is equivalent to

$$(29.14) \qquad T_1 > t\,, \qquad T_2 > t\,, \qquad T_3 > t\,;$$

the first factor is therefore

$$B_1 = v_1(t).v_2(t).v_3(t)\,.$$

The second factor B_2 is a conditional probability, given that (29.14) occurs; but in this case, the conditional laws of the T_i may be written (see Section 7, survival law of nonnew equipment) as

$$(29.15) \qquad w_k(u) \equiv \mathrm{pr}\,\{\,T_k > u \mid T_k > t\,\} = 1\,, \qquad u \leqslant t\,,$$
$$= \frac{v_k(u)}{v_k(t)}\,, \qquad u \geqslant t\,.$$

It then follows that

(29.16)

$$B_2 = 1 - \mathrm{pr}\,\{\,T_3 > T_1 \mid \min\,(T_1,\,T_2,\,T_3) > t\,\} = 1 - \int_t^\infty \frac{i_1(u)}{v_1(t)}\,\frac{v_3(u)}{v_3(t)}\,du\,,$$

from which

$$B = B_1\,.\,B_2 = v_1(t)\,v_2(t)\,v_3(t) - v_2(t)\int_t^\infty i_1(u)\,v_3(u)\,du\,,$$

and finally

$$(29.17) \qquad v(t) = v_1(t)\,v_2(t)\,v_3(t) + v_2(t)\int_0^t i_1(u)\,v_3(u)\,du\,.$$

By differentiating this expression and performing an integration by parts in order to transform the integral that appears in the derivative, one may easily verify that the probability density (29.6) is indeed recovered. On the other hand, in the particular case of exponential laws ((29.7) and (29.8)), (29.10) can be developed after a few simple transformations.

To finish, we indicate the mean lifetime of the system in the particular case of exponential laws:

$$(29.18) \qquad \overline{T} = \frac{\lambda_1}{\lambda_1 + \lambda_3}\,\frac{1}{\lambda_2} + \frac{\lambda_3}{\lambda_1 + \lambda_3}\,\frac{1}{\lambda_1 + \lambda_2 + \lambda_3}\,.$$

Case of Monotone Structures. We have seen at the beginning of this section that, if a system has a monotone structure, since the probability that it fails for the first time at the instant t is the probability that it is in a good state at time t, we have

$$(29.19) \qquad v(t) = h[v_1(t),\,v_2(t),\,\ldots,\,v_n(t)]$$

where $v(t)$, $v_1(t)$, ..., $v_n(t)$ are, respectively, the survival functions for the system and for its components, and where $h(p_1, p_2, \ldots, p_n)$ is the reliability function (25.7), which itself may be obtained from the structure function $\varphi(x)$ (see Eq. (25.9)). In order to obtain the survival function of a monotone system, it therefore suffices to substitute in the simple form of the structure function for the variables x_i, the survival functions $v_i(t)$ of its components.

Example 1. Consider again the structure defined in Fig. 25.1. First, for the structure function, we have, according to (25.11) and (25.12),

$$\varphi(x) = 1 - (1 - x_2 x_5)(1 - x_2 x_6)(1 - x_1 x_4 x_5) \tag{29.20}$$
$$= x_2 x_5 + x_2 x_6 + x_1 x_4 x_5 - x_2 x_5 x_6 - x_1 x_2 x_4 x_5 .$$

From this, the reliability function

$$h(p) = p_2 p_5 + p_2 p_6 + p_1 p_4 p_5 - p_2 p_5 p_6 - p_1 p_2 p_4 p_5 , \tag{29.21}$$

and the survival function

$$v(t) = v_2(t).v_5(t) + v_2(t).v_6(t) + v_1(t).v_4(t).v_5(t) \tag{29.22}$$
$$- v_2(t).v_5(t).v_6(t) - v_1(t).v_2(t).v_4(t).v_5(t) .$$

In the particular case where

$$v_6(t) = v_5(t) = v_4(t) = v_3(t) = v_2(t) = v_1(t) , \tag{29.23}$$

we have

$$v(t) = 2\, v_1^2(t) - v_1^4(t) . \tag{29.24}$$

Example 2. Consider the bridge network of Fig. 19.10, p. 79. This network has three minimal links, which are

$$\{ e_1, e_2 \} , \qquad \{ e_1, e_3 \} , \qquad \{ e_1, e_4 \} . \tag{29.25}$$

Formula (22.1) gives

$$\varphi(x) = 1 - (1 - x_1 x_2)(1 - x_1 x_3)(1 - x_1 x_4) , \tag{29.26}$$

from which

(29.27)

$$\varphi_s(x) = x_1 x_2 + x_1 x_3 + x_1 x_4 - x_1 x_2 x_3 - x_1 x_2 x_4 - x_1 x_3 x_4 + x_1 x_2 x_3 x_4$$

and

(29.28)

$$v(t) = v_1(t)\, [v_2(t) + v_3(t) + v_4(t) - v_2(t)\, v_3(t) - v_2(t)\, v_4(t) - v_3(t)\, v_4(t)$$
$$+ v_2(t)\, v_3(t)\, v_4(t)] .$$

Duality. In Section 17, we have defined for any structure function $\varphi(x)$, a dual structure function $\overline{\varphi}(x)$:

$$\overline{\varphi}(x) = 1 - \varphi(1 - x) \tag{29.29}$$

or, in a more detailed fashion:

$$\overline{\varphi}(x_1, x_2, \ldots, x_n) = 1 - \varphi[(1 - x_1), (1 - x_2), \ldots, (1 - x_n)] . \tag{29.30}$$

To (29.30) corresponds a duality relation for the corresponding reliability functions

$$\overline{h}(p) = 1 - h(1 - p) \tag{29.31}$$

or, in a more detailed fashion,

$$\overline{h}(p_1, p_2, \ldots, p_n) = 1 - h[(1 - p_1), (1 - p_2), \ldots, (1 - p_n)] . \tag{29.32}$$

From this we have a duality relation for the survival functions

$$\begin{aligned} \overline{v}(t) &= \overline{h}[v_1(t), v_2(t), \ldots, v_n(t)] \\ &= 1 - h[(1 - v_1(t)), (1 - v_2(t)), \ldots, (1 - v_n(t))] \\ &= 1 - h[\Phi_1(t), \Phi_2(t), \ldots, \Phi_n(t)] , \end{aligned} \tag{29.33}$$

where

$$\Phi_i(t) = 1 - v_i(t) , \quad i = 1, 2, \ldots, n . \tag{29.34}$$

Example. Consider two components e_1 and e_2 for which we have the structure function

$$\varphi(x) = x_1 . x_2 . \tag{29.35}$$

From this

$$\overline{\varphi}(x) = 1 - (1 - x_1)(1 - x_2) , \tag{29.36}$$

$$\overline{h}(p) = 1 - (1 - p_1)(1 - p_2) , \tag{29.37}$$

and

$$\begin{aligned} \overline{v}(t) &= 1 - (1 - v_1(t))(1 - v_2(t)) \\ &= 1 - \Phi_1(t) . \Phi_2(t) . \end{aligned} \tag{29.38}$$

Suppose that

$$v_1(t) = v_2(t) = e^{-\lambda t} ; \tag{29.39}$$

we have

$$v(t) = v_1(t) . v_2(t) = e^{-2\lambda t} \tag{29.40}$$

and

$$\begin{aligned} \bar{v}(t) &= 1 - (1 - v_1(t))\,(1 - v_2(t)) \\ &= 1 - (1 - v_1(t))^2 \\ &= 2\,v_1(t) - v_1^2(t) \\ &= 2\,e^{-\lambda t} - e^{-2\lambda t} \\ &= e^{-\lambda t}(2 - e^{-\lambda t})\,. \end{aligned} \tag{29.41}$$

Monotone Linear Composition of Survival Functions. We have seen in Section 26 (Theorem 26.V, p. 129) that a monotone structure function $\varphi(x_1, x_2, \ldots, x_n)$ may always be put in the form

$$\varphi(x_1, x_2, \ldots, x_n) = x_n\,\varphi_A(x_1, \ldots, x_{n-1}) + (1 - x_n)\,\varphi_B(x_1, \ldots, x_{n-1})\,, \tag{29.42}$$

where

$$\varphi_A(x_1, \ldots, x_{n-1}) \geqslant \varphi_B(x_1, \ldots, x_{n-1}) \tag{29.43}$$

for any $(x_1, \ldots, x_{n-1})$, $x_i = 0$ or 1.

If φ_A and φ_B are in simple form, expression (29.42) is linear with respect to each of the variables, and, if one designates by $p_1, p_2, \ldots, p_n$ the reliabilities of the components at an arbitrary instant, the reliability functions of the system S and its substructures A and B are related in the same way, according to (26.53):

$$\varphi(p_1, p_2, \ldots, p_n) = p_n\,\varphi_A(p_1, \ldots, p_{n-1}) + (1 - p_n)\,\varphi_B(p_1, \ldots, p_{n-1}) \tag{29.44}$$

with

$$\varphi_A(p_1, \ldots, p_{n-1}) \geqslant \varphi_B(p_1, \ldots, p_{n-1}) \tag{29.45}$$

for all

$$(p_1, \ldots, p_{n-1})\,, \quad 0 \leqslant p_i \leqslant 1\,, \tag{29.46}$$

and

$$\varphi \equiv h\,, \qquad \varphi_A \equiv h_A\,, \qquad \varphi_B \equiv h_B\,. \tag{29.47}$$

Note that (29.43), which is valid for $x_i = 0$ or 1, implies (29.45) because of the linearity of the functions φ, φ_A, and φ_B.

Relation (29.19) then shows that the survival function of the system S, and those of its substructures, may be obtained simply by replacing in (29.44) the reliabilities p_i of the components by their values $v_i(t)$:

$$v(t) = v_n(t) . v_A(t) + [1 - v_n(t)]\,v_B(t) \tag{29.48}$$

where

$$v_A(t) = \varphi_A[v_1(t), v_2(t), \ldots, v_{n-1}(t)], \tag{29.49}$$

$$v_B(t) = \varphi_B[v_1(t), v_2(t), \ldots, v_{n-1}(t)], \tag{29.50}$$

with

$$v_A(t) \geqslant v_B(t), \qquad \forall t \geqslant 0. \tag{29.51}$$

Relation (29.48) permits us to study the lifetime T of a system S considered as formed through monotone linear composition of two subsystems A and B, as a function of the respective lifetimes T_A, T_B, and T_n of these subsystems and of a supplementary component e_n.

Thus the mean lifetime $\overline{T}$, given from the survival function $v(t)$ by (5.11),

$$\overline{T} = \int_0^\infty v(t)\,dt, \tag{29.52}$$

may be written, according to (29.48), as

$$\overline{T} = \int_0^\infty v_n(t)\,v_A(t)\,dt + \int_0^\infty [1 - v_n(t)]\,v_B(t)\,dt \tag{29.53}$$

or

$$\overline{T} = \overline{T}_B + \int_0^\infty v_n(t)\,[v_A(t) - v_B(t)]\,dt, \tag{29.54}$$

by noting that

$$\int_0^\infty v_B(t)\,dt = \overline{T}_B. \tag{29.55}$$

Inequality (29.51) shows that

$$\overline{T} \geqslant \overline{T}_B. \tag{29.56}$$

As one might expect, the mean lifetime of a system is at least equal to that of the weakest subsystem (evidently, $\overline{T}_B \leqslant \overline{T}_A$).

Setting

$$\Phi_n(t) = 1 - v_n(t) = \text{pr}\,\{ T_n \leqslant t \}, \tag{29.57}$$

we may also write

$$\overline{T} = \overline{T}_A - \int_0^\infty \Phi_n(t)\,[v_A(t) - v_B(t)]\,dt, \tag{29.58}$$

from which

$$\overline{T} \leqslant \overline{T}_A. \tag{29.59}$$

Therefore $\overline{T}$ is included between $\overline{T}_B$ and $\overline{T}_A$.

The cumulative failure rate of the system, defined by Eq. (4.18),

$$\Lambda(t) = \int_0^t \lambda(u)\,\mathrm{d}u\,, \tag{29.60}$$

where $\lambda(t)$ is the instantaneous failure rate (4.8)

$$\lambda(t) = -\frac{v'(t)}{v(t)} = \frac{i(t)}{v(t)}\,,$$

and is obtained from $v(t)$ through relation (4.20):

$$\Lambda(t) = -\ln v(t)\,. \tag{29.61}$$

From (29.48), we have

$$\Lambda(t) = -\ln\left[v_n(t).v_A(t) + v_B(t) - v_n(t).v_B(t)\right]. \tag{29.62}$$

The instantaneous failure rate is

$$\lambda(t) = \frac{i_n(t)\,v_A(t) + v_n(t)\,i_A(t) + i_B(t) - i_n(t)\,v_B(t) - v_n(t)\,i_B(t)}{v_n(t)\,v_A(t) + v_B(t) - v_n(t)\,v_B(t)}\,, \tag{29.63}$$

or, by suppressing to simplify the notation all arguments of the functions, all being t,

$$\lambda = \frac{\lambda_n\,v_n(v_A - v_B) + \lambda_A\,v_n\,v_A + \lambda_B\,v_B(1 - v_n)}{v_n\,v_A + v_B - v_n\,v_B}\,. \tag{29.64}$$

Example. Weibull functions with the same form coefficient. The above expressions generally require turning to a computer for numerical computation. We shall go on to see, however, by way of an example, a case where the mean lifetime may be expressed in a very simple fashion. We suppose that the lifetimes of the subsystems A and B and of the component e_n each follow a Weibull law (6.18):

$$v_k(t) = \mathrm{e}^{-(\alpha_k t)^\beta}\,, \qquad k = A, B, n\,, \tag{29.65}$$

the form coefficient β being the same in all three cases.

Relation (29.53) then gives

(29.66)

$$\overline{T} = \int_0^\infty \mathrm{e}^{-(\alpha_n t)^\beta}.\mathrm{e}^{-(\alpha_A t)^\beta}\,\mathrm{d}t + \int_0^\infty \mathrm{e}^{-(\alpha_B t)^\beta}\,\mathrm{d}t - \int_0^\infty \mathrm{e}^{-(\alpha_n t)^\beta}.\mathrm{e}^{-(\alpha_B t)^\beta}\,\mathrm{d}t\,.$$

Set

$$a = (\alpha_n^\beta + \alpha_A^\beta)^{1/\beta}\,, \tag{29.67}$$

$$b = (\alpha_n^\beta + \alpha_B^\beta)^{1/\beta}\,. \tag{29.68}$$

It follows that

$$\bar{T} = \int_0^\infty e^{-(at)^\beta}\,dt + \int_0^\infty e^{-(\alpha_B t)^\beta}\,dt - \int_0^\infty e^{-(bt)^\beta}\,dt\,. \tag{29.69}$$

According to (6.21) we have

$$\int_0^\infty e^{-(\alpha t)^\beta}\,dt = \frac{1}{\alpha}\Gamma\left(1 + \frac{1}{\beta}\right), \tag{29.70}$$

from which

$$\bar{T} = \left(\frac{1}{\alpha_B} + \frac{1}{a} - \frac{1}{b}\right).\Gamma\left(1 + \frac{1}{\beta}\right). \tag{29.71}$$

In the particular case where $\beta = 1$, that is, when the three survival laws of A, B, and e_n are exponential laws, we obtain

$$\bar{T} = \frac{1}{\lambda_B} + \frac{1}{\lambda_n + \lambda_A} - \frac{1}{\lambda_n + \lambda_B}, \tag{29.72}$$

where $\lambda_n = \alpha_n$, $\lambda_A = \alpha_A$, and $\lambda_B = \alpha_B$ are the respective failure rates of e_n, A, and B. Note that in this case

$$v(t) = e^{-\lambda_B t} + e^{-(\lambda_n + \lambda_A)t} - e^{-(\lambda_n + \lambda_B)t}\,. \tag{29.73}$$

IFRA Survival Functions. We shall see in Section 30 that a system with monotone structure function having components with IFR survival functions (increasing failure rate, see Sections 10 and 11) does not necessarily have an IFR function itself. On the contrary, the survival function of the system is IFRA (increasing failure rate average, see Section 12); it even suffices that the components have an IFRA survival function, as the following theorem shows [9].

Theorem 29.I. *A system with monotone structure function having components with IFRA survival functions similarly has an IFRA survival function.*

We prove this theorem by recurrence using monotone linear composition (see Section 26). Recall that a survival function $v(t)$ is IFRA if the function

$$L(t) = \Lambda(t)/t \tag{29.74}$$

is nondecreasing for all $t \geqslant 0$. We then have

$$L'(t) = \frac{t\lambda(t) - \Lambda(t)}{t^2} \geqslant 0\,, \tag{29.75}$$

that is,

$$t\lambda(t) \geqslant \Lambda(t)\,. \tag{29.76}$$

Let $e_1, e_2, \ldots, e_n$ be the components of the system; decompose the system into two substructures A and B by isolating the component e_n (see (29.42)); and suppose that the survival functions $v_A(t)$, $v_B(t)$, and $v_n(t)$ of A, B, and e_n are IFRA. We then have

$$t\lambda_k(t) \geqslant \Lambda_k(t)\,, \qquad k = A, B, n\,. \tag{29.77}$$

Using expression (29.64) for the failure rate $\lambda(t)$ of the system and the relation

$$\Lambda_k(t) = -\ln v_k(t)\,, \tag{29.78}$$

we may see that relations (29.77) imply that

$$t\lambda(t) \geqslant \frac{\Lambda_n v_n(v_A - v_B) + \Lambda_A v_n v_A + \Lambda_B v_B(1 - v_n)}{v_n v_A + v_B - v_n v_B}\,, \tag{29.79}$$

or

$$t\lambda(t) \geqslant -\frac{v_n(v_A - v_B)\ln v_n + v_n v_A \ln v_A + v_B(1 - v_n)\ln v_B}{v_n v_A + v_B - v_n v_B}\,. \tag{29.80}$$

To show that $v(t)$ is IFRA, that is, to verify (29.76), it suffices to show that

$$\frac{v_n(v_A - v_B)\ln v_n + v_n v_A \ln v_A + v_B(1 - v_n)\ln v_B}{v_n v_A + v_B - v_n v_B} \geqslant \Lambda(t) = -\ln(v_n v_A + v_B - v_n v_B)\,, \tag{29.81}$$

for all values of v_n, v_A, and v_B satisfying the conditions

$$0 \leqslant v_n \leqslant 1 \quad \text{and} \quad 0 \leqslant v_B \leqslant v_A \leqslant 1\,, \tag{29.82}$$

which express that we have survival functions and that the system is obtained through monotone linear composition (see (29.51)).

To simplify the notation, set

$$a = v_A\,, \qquad b = v_B\,, \qquad c = v_n\,, \tag{29.83}$$

with

$$0 \leqslant c \leqslant 1 \text{ and } 0 \leqslant b \leqslant a \leqslant 1\,. \tag{29.84}$$

The inequality to be shown under the conditions in (29.84) may then be written as

$$c(a-b)\ln c + ca\ln a + (1-c)\,b\ln b \leqslant [ca + (1-c)\,b]\ln[ca + (1-c)\,b]\,, \tag{29.85}$$

or, by setting

$$f(x) = x\ln x\,, \tag{29.86}$$

as

$$(a - b)\,f(c) + cf(a) + (1 - c)\,f(b) \leqslant f[ca + (1 - c)\,b]\,. \tag{29.87}$$

The function $f(x)$ is convex and negative for $0 < x < 1$ (see Fig. 29.2); the difference

$$cf(a) + (1 - c) f(b) - f[ca + (1 - c) b]$$

represents the length of the segment MN, the distance along the ordinates between the chord RS and the curve $f(x)$ at the point $ca + (1 - c)b$ on the abscissa; it is then a matter of showing that the segment is at most equal to the product of the segments AB and PQ, where Q divides the segment (0, 1) as T divides the segment AB.

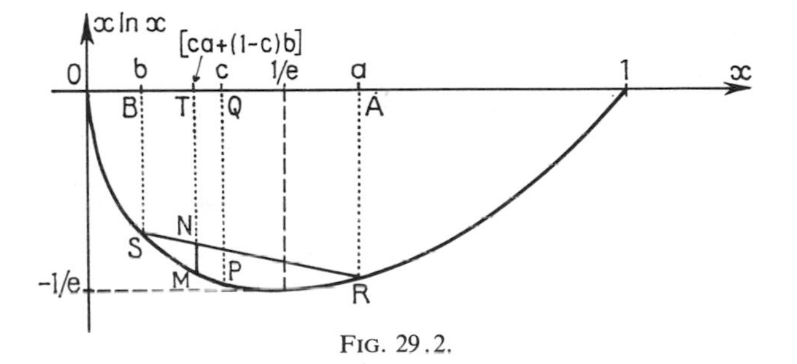

FIG. 29.2.

For $b = 0$, relation (29.87) becomes

$$af(c) + cf(a) \leqslant f(ca)\,, \tag{29.88}$$

thus

$$ac \ln c + ac \ln a \leqslant ac \ln (ac)\,, \tag{29.89}$$

a relation that is evidently satisfied (with equality and not with strict inequality). For a and c fixed, put

$$g(b) = f[ca + (1 - c) b] - cf(a) - (1 - c) f(b) - (a - b) f(c)\,. \tag{29.90}$$

We just saw that $g(0) = 0$, and it is to be shown that $g(b) \geqslant 0$ for $0 \leqslant b \leqslant a$, and this is to hold for any a and c in the segment (0, 1). We have, however,

$$g(b) = [ca + (1 - c) b] \ln [ca + (1 - c) b] - ca \ln a - (1 - c) b \ln b - c(a - b) \ln c\,, \tag{29.91}$$

$$g'(b) = c \ln c - (1 - c) \ln b + (1 - c) \ln [ac + (1 - c) b]\,, \tag{29.92}$$

$$g''(b) = -\frac{ac(1 - c)}{b[ca + (1 - c) b]} < 0\,. \tag{29.93}$$

From (29.93) the function $g(b)$ is concave; it thus attains its minimum in the interval $(0, a)$ either for $b = 0$ or $b = a$; we have seen that $g(0) = 0$, and similarly

$$g(a) = a \ln a - ca \ln a - (1 - c) a \ln a = 0\,. \tag{29.94}$$

It then follows that $g(b) \geqslant 0$ for $0 \leqslant b \leqslant a$, which is what we set out to prove.

We have thus seen that if a system is formed through linear composition of two subsystems A and B with a component e_n such that the survival functions v_A, v_B, and v_n are IFRA, the survival function v of the system is also IFRA.

Given a system satisfying the hypotheses of the theorem, we may decompose it into two subsystems by isolating one component, chosen arbitrarily, for example, e_n. If the subsystems obtained are of order greater than 1, we may in turn decompose these by the same process, and continue thusly until subsystems of order 0 or 1 are obtained, that is, components (with structure function $\varphi_1^{(1)}(x_i) = x_i$; see Fig. 26.13, p. 134) or degenerate structures ($\varphi_1^{(0)} = 1$ or $\varphi_2^{(0)} = 0$). However, a degenerate system has survival function $v(t) \equiv 0$ or $v(t) \equiv 1$, both of which are trivially IFRA, and a system of order 1 evidently satisfies the theorem. We have therefore proved the theorem.

Thus for monotone systems with IFRA components we may use the properties proved in Section 12. In particular, according to the corollary to Theorem 12.V, p. 53, the coefficient of variation $\sigma_T/\overline{T}$ of the lifetime of such a system is greater than or equal to 1, that is, T may not have a variance greater than that corresponding to an exponential law.

On the other hand, if we have calculated for a certain value of t, the value of the survival function $v(t)$ of the system using relation (29.19), we may then deduce the cumulative failure rate

$$\Lambda(t) = -\ln v(t) \tag{29.95}$$

and the mean instantaneous failure rate between 0 and t (see (12.4))

$$L(t) = \frac{1}{t}\int_0^t \lambda(u)\, du = \frac{\Lambda(t)}{t}. \tag{29.96}$$

If we have to evaluate the reliability of a system for a period of use τ near to t, we may use the approximation

$$v(\tau) \simeq v_e(\tau) = e^{-\bar{\lambda}\tau} \quad \text{for} \quad \tau \simeq t \tag{29.97}$$

where $\bar{\lambda} = L(t)$. We are here reasoning as if the system had a constant failure rate, equal to $\bar{\lambda}$. Theorem 12.I (p. 47) shows that by so reasoning, the system reliability is underestimated if $\tau < t$, and is overestimated if $\tau > t$, since

$$\begin{aligned} v(\tau) &\geqslant v_e(\tau), \qquad 0 < \tau < t, \\ v(\tau) &\leqslant v_e(\tau), \qquad \tau > t. \end{aligned} \tag{29.98}$$

The "mean" failure rate $\bar{\lambda} = L(t)$ will, of course, depend on the period of use t taken as a reference. More precisely, it is a nondecreasing function of

t, by the very definition of the IFRA property for a system. It is therefore between $L(0)$ and $L(\infty)$, but

$$L(0) = \lim_{t \to 0} \frac{1}{t} \int_0^t \lambda(u)\, du = \lambda(0) \tag{29.99}$$

is the failure rate of the system at the origin. One may show by using expression (22.5) for the structure function and the notion of random structure function (Section 25) that:

(1) the failure rate at the origin is zero if the width of the structure is greater than 1 (see Section 17, p. 66); that is, there does not exist a cut of just one component. Indeed, in this case failure of the system involves failure of at least two components, an event whose probability is of second order for $t \to 0$, by virtue of the independence of the lifetimes of the components (hypothesis (4) of Section 24, p. 115);

(2) if there exist cuts of just one component, the failure rate of the system at the origin is the sum of the failure rates at the origin of the components that occur in these cuts.

Calculation of the superior limit $L(\infty)$ of the "mean" failure rate is more delicate. Esary and co-workers [20] have shown that, in the particular case of all components having an exponential lifetime, and therefore a constant failure rate, the limit for $t \to \infty$ of the mean failure rate $L(t)$ is equal to the minimum, calculated on the set of links of the structure, of the sum of the failure rates of all the components occurring in a link.

To conclude, we note that it is the "mean" failure rate $L(t)$ of the system that is in the interval $[L(0), L(\infty)]$; the instantaneous failure rate of the system may, for some values of t, exceed the value $L(\infty)$ since the system is IFRA and not, in general, IFR.

30 Survival Functions for Series and Parallel Structures. Asymptotic Results for a Large Number of Components

Series Structure. Consider a series structure of n components; its structure function is

$$\varphi(x) = x_1 x_2 \ldots x_n, \tag{30.1}$$

the reliability function is

$$h(p) = \varphi(p) = p_1 p_2 \cdots p_n, \tag{30.2}$$

and the survival function is

$$v(t) = h(v_1(t), v_2(t), \ldots, v_n(t)) = v_1(t) . v_2(t) . \ldots . v_n(t) \tag{30.3}$$

where $v_1(t), v_2(t), \ldots, v_n(t)$ are the survival functions of its components. This relation may also be written as

$$\text{Log}\, v(t) = \text{Log}\, v_1(t) + \text{Log}\, v_2(t) + \cdots + \text{Log}\, v_n(t) . \tag{30.4}$$

It follows that the cumulative failure rate $\Lambda(t)$ of the system (cf. Section 4, or (29.60) and (29.61)) is

$$\Lambda(t) = \sum_{i=1}^{n} \Lambda_i(t) , \tag{30.5}$$

from which it follows by differentiation that

$$\lambda(t) = \sum_{i=1}^{n} \lambda_i(t) . \tag{30.6}$$

The failure rate (instantaneous or cumulative) of the system is therefore the sum of the failure rates of its components. As a consequence, if all the components have an IFR survival function (respectively, DFR; see Sections 10 and 11), the system similarly has an IFR (respectively, DFR) survival function.

In particular, if all the components have an exponential survival function

$$v_i(t) = e^{-\lambda_i t} \tag{30.7}$$

with a constant failure rate $\lambda_i(t) = \lambda_i$, then it is the same for the system:

$$v(t) = e^{-\lambda t} \tag{30.8}$$

where λ is given by (30.6). The series structure thus presents particularly simple properties. It is at the same time the most common; it corresponds to the case where the failure of any single component entails failure of the system.

Parallel Structure. Consider a parallel structure of n components; the structure function is

$$\varphi(x) = 1 - \prod_{i=1}^{n} (1 - x_i) , \tag{30.9}$$

and the survival function is

$$v(t) = 1 - \prod_{i=1}^{n} (1 - v_i(t)) . \tag{30.10}$$

Let $\Phi(t) = 1 - v(t)$ be the distribution law of the lifetime of the structure, and $\Phi_i(t) = 1 - v_i(t)$ be the distribution of the lifetime of component i. We obtain

$$\Phi(t) = \prod_{i=1}^{n} \Phi_i(t) . \tag{30.11}$$

We now calculate the failure rate $\lambda(t)$ of the structure

$$(30.12)\qquad \lambda(t) = -\frac{v'(t)}{v(t)} = \frac{\Phi'(t)}{v(t)} = \frac{\sum_{i=1}^{n} \frac{\Phi_i'(t)}{\Phi_i(t)} \Phi(t)}{v(t)} .$$

By noting that

$$(30.13)\qquad \frac{\Phi_i'(t)}{\Phi_i(t)} = -\frac{v_i'(t)}{1 - v_i(t)} = \frac{v_i(t)}{1 - v_i(t)} \lambda_i(t)$$

we obtain

$$(30.14)\qquad \frac{v(t)}{1 - v(t)} \lambda(t) = \sum_{i=1}^{n} \frac{v_i(t)}{1 - v_i(t)} \lambda_i(t) .$$

If the components have exponential survival laws, $\lambda_i(t) = \lambda_i$,

$$(30.15)\qquad \frac{v(t)}{1 - v(t)} \lambda(t) = \sum_{i=1}^{n} \frac{e^{-\lambda_i t}}{1 - e^{-\lambda_i t}} \lambda_i$$

with

$$(30.16)\qquad v(t) = 1 - \prod_{i=1}^{n} (1 - e^{-\lambda_i t}) .$$

Remark. If the components of a system S have an exponential survival law and S does not have a series structure, then the survival function of S is not exponential. Indeed, we shall see in Example 1 below that the failure rate given by (30.15) is not monotone, which shows that a system with monotone structure having components with IFR survival functions (which is the case with the exponential law) does not necessarily have an IFR survival function (we have seen however in Theorem 29.I that it is IFRA).

Examples.

(1) Consider a system composed of two components in parallel, for which the survival laws are exponential:

$$(30.17)\qquad v(t) = 1 - (1 - e^{-\lambda_1 t})(1 - e^{-\lambda_2 t}) = e^{-\lambda_1 t} + e^{-\lambda_2 t} - e^{-(\lambda_1 + \lambda_2)t} ,$$

$$(30.18)\qquad \lambda(t) = -\frac{v'(t)}{v(t)} = \frac{\lambda_1 e^{-\lambda_1 t} + \lambda_2 e^{-\lambda_2 t} - (\lambda_1 + \lambda_2) e^{-(\lambda_1 + \lambda_2)t}}{e^{-\lambda_1 t} + e^{-\lambda_2 t} - e^{-(\lambda_1 + \lambda_2)t}} .$$

Set $\lambda_1 + \lambda_2 = 1$ and take different values for λ_1/λ_2; we obtain the graph of Fig. 30.1.

The failure rate tends, for very large t, toward a limit that is equal to the failure rate of the best of the components.

(2) In order to increase reliability, one places in parallel with a piece of equipment A a second piece of identical equipment B. A switch C automatically assures exchange of the equipment to be used in case of failure.

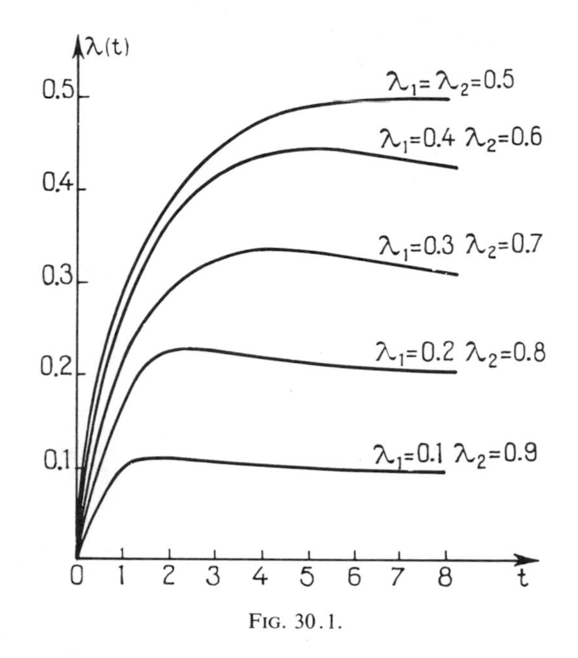

FIG. 30.1.

We suppose that the fact of being used or not does not influence the rate of failure for the equipment[7] and that any failure of the switch interrupts functioning of the system. We shall examine whether this "redundance" in fact increases the reliability, as we would hope.

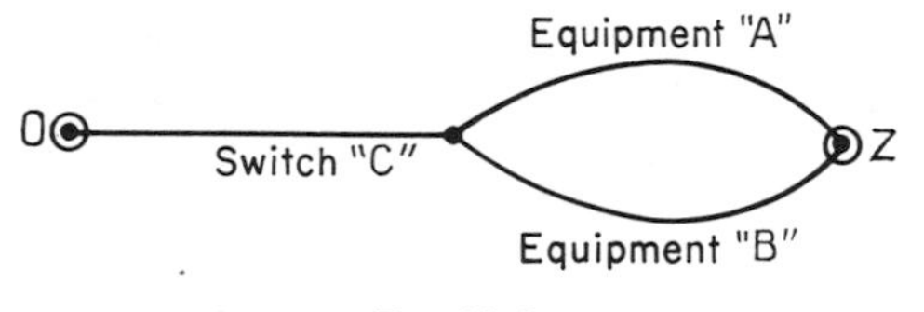

FIG. 30.2.

The reliability network is shown in Fig. 30.2; it is a series–parallel structure, C being in series with the two pieces of equipment A and B in parallel. The links are $\{ C, A \}$ and $\{ C, B \}$; the structure function is therefore (cf. Eq. (22.1))

$$\begin{aligned} \varphi(x) &= 1 - (1 - x_C\, x_A)\,(1 - x_C\, x_B) \\ &= x_C(x_A + x_B - x_A\, x_B) \end{aligned} \tag{30.19}$$

[7] If this hypothesis is not made, then the probability of failure of one piece of equipment would depend on the state of the other. This would contradict hypothesis (4) of Section 24, p. 115, and the theory developed here would not be applicable; see also Chapter V, Section 32.

(since $x_C^2 = x_C$) and the reliability function is

$$h(p) = p_C(2\,p_E - p_E^2) \tag{30.20}$$

where $h(p)$ is the reliability of the system; p_C is the reliability of the switch; and p_E is the reliability of the equipment (A or B).

Redundance is therefore useful if

$$h(p) > p_E \tag{30.21}$$

or

$$p_C(2 - p_E) > 1\,. \tag{30.22}$$

In Fig. 30.3 we have traced the curve $p_C(2 - p_E) = 1$; this defines two zones:

upper zone: redundance is useful,
lower zone: redundance is detrimental.

If we can vary the time of functioning t, the point with coordinates $\{ p_E(t), p_C(t) \}$ will describe a curve like the dashed curve in the figure. If t is small ($t \to 0$), p_C and p_E are near to 1 and the redundance will be useful if $p_C > p_E$ (the reliability of the switch is greater than the reliability of the equipment). If t is large ($t \to \infty$), p_C and p_E are small, and the redundance will always be detrimental.

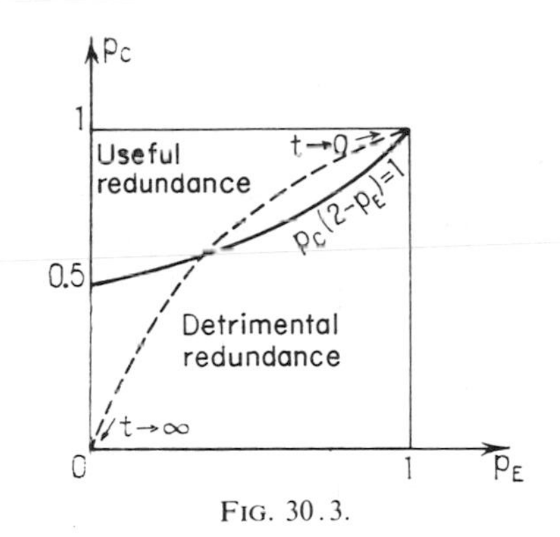

FIG. 30.3.

Study of Systems Composed of a Large Number of Identical Components. Consider a system composed of n components and let $T_1, T_2, \ldots, T_n$ be the lifetimes of these components. If all the components are in series, the first failure of a component will entail failure of the system, and the lifetime of the system is

$$T_S = \min(T_1, T_2, \ldots, T_n)\,. \tag{30.23}$$

If, on the other hand, all the components are in parallel, the system lifetime will be

$$T_P = \max (T_1, T_2, \ldots, T_n) . \tag{30.24}$$

If the number n of components is very large, the distribution law of T_S and T_P is studied by the theory of extreme values (cf. Gumbel [26]).

We present here, in a simplified way, certain results of this theory allowing applications to the study of reliability.

The survival function $v_S(t)$ of a system of n series components is

$$v_S(t) = v(t)^n \tag{30.25}$$

where $v(t)$ is the survival function of each component. If n tends toward infinity and if $v(t) < 1$, then $v_S(t)$ tends toward 0.

We shall concern ourselves only with the case where $v(t)$ is near to 1, that is, t sufficiently small ($t \to 0$) in order that the components have excellent reliabilities. Then

$$v_S(t) = e^{n \operatorname{Log} v(t)} \sim e^{n[1 - v(t)]} . \tag{30.26}$$

The survival function $v_P(t)$ of a system of n parallel components is

$$v_P(t) = 1 - (1 - v(t))^n . \tag{30.27}$$

If n tends toward infinity and if $v(t) > 0$, then $v_P(t)$ tends toward 1. We restrict ourselves to the case where $v(t)$ is near 0 (t sufficiently large so that the components have very weak reliabilities). Then

$$v_P(t) = 1 - e^{n \operatorname{Log} [1 - v(t)]} \sim 1 - e^{-nv(t)} . \tag{30.28}$$

Examples.

(1) Let

$$v(t) = \begin{cases} 1 \quad \text{if} \quad t \leqslant t_0 , \\ 1 - \beta(t - t_0)^\alpha [1 + \varepsilon(t - t_0)] \quad \text{if} \quad t_0 \leqslant t \leqslant t_1 , \\ 0 \quad \text{if} \quad t \geqslant t_1 , \end{cases} \tag{30.29}$$

where t_1 is defined by

$$1 - \beta(t_1 - t_0)^\alpha [1 + \varepsilon(t_1 - t_0)] = 0 , \tag{30.30}$$

and where $\varepsilon(t - t_0)$ is a function that tends toward 0 when t tends toward t_0. This survival law has the form given in Fig. 30.4.

We have for $t \to t_0$ ($t > t_0$),

$$\operatorname{Log} v(t) \sim - \beta(t - t_0)^\alpha . \tag{30.31}$$

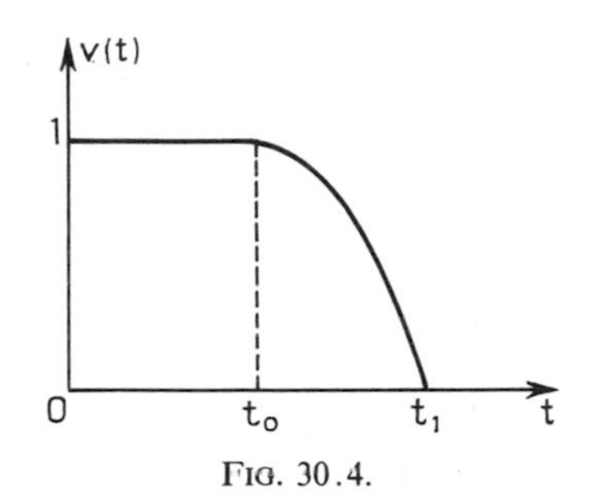

FIG. 30.4.

The survival law of a system of n components in series is thus

$$(30.32)\qquad v_S(t)\begin{cases} = 1 & \text{if} \quad t \leqslant t_0\,, \\ \sim e^{-n\beta(t-t_0)^\alpha} & \text{if} \quad t \leqslant t_1,\, t \to t_1\,. \end{cases}$$

Therefore $v_S(t)$ is near a Weibull law ((Eq. 6.25′)) for $t \to t_1$. If $t_0 = 0$ and $\alpha = 1$, we obtain the exponential law (cf. Eq. (30.8))

$$(30.33)\qquad v_S(t) \sim e^{-n\beta t}\,.$$

(2) If

$$(30.34)\qquad v(t) \sim e^{-\lambda t^k} \quad \text{as} \quad t \to \infty$$

we obtain for n components in parallel,

$$(30.35)\qquad v_P(t) \sim 1 - e^{-n e^{-\lambda t^k}}$$

and, in the particular case where $k = 1$, we have

$$(30.36)\qquad v(t) \sim e^{-\lambda t}\,, \qquad v_P(t) \sim 1 - e^{-n e^{-\lambda t}}\,.$$

(3) If

$$(30.37)\qquad v(t) \sim \frac{a}{t^\alpha} \quad \text{as} \quad t \to \infty\,, \qquad a, \alpha \text{ positive},$$

we obtain for n components in parallel,

$$(30.38)\qquad v_P(t) \sim 1 - \exp\left(-\,n\frac{a}{t^\alpha}\right) \qquad (t \to \infty)\,.$$

We give two particular cases of this formula:

$$(30.39)\qquad v(t) = \begin{cases} 1 & \text{if} \quad t \leqslant 1 \\ 1/t & \text{if} \quad t \geqslant 1 \end{cases} \Rightarrow v_P(t)\begin{cases} = 1 & \text{if} \quad t \leqslant 1 \\ \sim 1 - e^{-n/t} & \text{if} \quad t \gg 1; \end{cases}$$

$$(30.40)\qquad v(t) = \begin{cases} 1 & \text{if} \quad t \leqslant 1 \\ 1/t^\alpha & \text{if} \quad t \geqslant 1 \end{cases} \Rightarrow v_P(t)\begin{cases} = 1 & \text{if} \quad t \leqslant 1 \\ \sim 1 - e^{-n/t^\alpha} & \text{if} \quad t \gg 1\,. \end{cases}$$

(4) Let

$$v(t) = \begin{cases} 0 & \text{if} \quad t \geqslant t_1 \\ \beta(t_1 - t)^\alpha [1 + \varepsilon(t_1 - t)] & \text{if} \quad 0 \leqslant t \leqslant t_1 \end{cases} \tag{30.41}$$

where $\varepsilon(t_1 - t)$ tends toward 0 if $t_1 - t$ tends toward 0, and where α and β are positive.

We obtain for n components in parallel

$$v_P(t) \begin{cases} = 0 & \text{if} \quad t \geqslant t_1 \\ \sim 1 - e^{-n\beta(t_1 - t)^\alpha} & \text{if} \quad t < t_1, \qquad t \to t_1 . \end{cases} \tag{30.42}$$

This is, afresh, a Weibull law.

CHAPTER V

REDUNDANCE

31 Introduction. Definitions

The term "redundance" is a general one which is used whenever one piece of equipment or a component may replace other equipment or another component that has failed. Any structure other than a series structure may be considered as redundant in the sense that there may be some failed components without having the system cease to be useful. On the contrary, a series structure has no redundance.

In a more particular fashion, redundance is a technique for increasing reliability, consisting of arranging in parallel two or more components, substructures (cf. Section 26), or complete devices. The theory of reliability of systems, developed in Chapters III and IV, permits the evaluation in each case of the reliability that is obtained by proceeding thus when designing a system.

Two reasons have, however, led us to devote a separate chapter to redundance. The first is that it poses some optimization problems for which general results or algorithms are available. The second derives from the fact that two types of redundance can be distinguished:

(a) Active Redundance. In the case of active redundance, all components of the system function permanently even if they are not strictly necessary, in the current state of the set of components. The hypotheses of Chapter IV (see Section 24) may then be admitted, in particular hypothesis

(4), according to which the lifetimes of the components are mutually independent in probability. This supposes, however, that each component has the same reliability, whether it alone assumes a given role, or when others share a part of its burden.

Examples.

(1) On certain trucks, two wheels are placed on each side of the rear axle; thus, in case of a flat, the other can assure support until assistance can be had. However, it will then be subject to greater stress, and its potential lifetime becomes considerably shorter. The hypothesis reviewed above is thus hardly realistic in this case.

(2) In an airplane engine, the ignition system is doubled; this increases the quality of ignition, and therefore the efficiency of the engine, but above all, in case of failure of one ignition system, the other alone can assure the functioning of the engine in an acceptable manner. One may reasonably admit in this case, if only because of the lack of precise data preventing one from making a better hypothesis, that the failure rate of the ignition system still in a good state is not modified by the failure of the other system.

We shall admit, in order to treat active redundance, that the hypotheses of Section 24 are satisfied. As we have indicated above, this will then make it possible for us to depend on the results of Chapters III and IV to study the optimization problems that arise. We shall see first (Section 32) that active redundance at the component level is more effective than that at the level of complex substructures; then we describe (Section 33) a method permitting, for a series structure, distribution of redundance in an optimal way by taking into account criteria of cost or size of the system. This method rests on a theoretical result, presented in Section 34, according to which the reliability of a monotone structure is a concave function of the number of redundant components. Finally, Section 35 will be dedicated to a particular kind of redundant structure, that said to be "k of n."

(b) Passive Redundance. Passive redundance consists in arranging some components or substructures as "in reserve," not to be used as long as no need is perceived. If an interruption in functioning of the system is tolerable, these elements will be manually put into service, by switching circuits or by replacing the failed elements; in the opposite case, automatic switching may be provided for.

Examples.

(1) The spare tire of an automobile constitutes a passive redundant element.

(2) The emergency generating equipment of a hospital is automatically activated in case of a power outage.

One generally supposes that unused elements are not subject to deterioration; the theory of Chapter IV is then not applicable since the failure rate of a redundant component depends on whether or not it has been placed in service (and, in the first case, on its age, which is different from that of the system), and as a consequence on the state of the other components of the system.

The study of passive redundance is very closely related to that of the maintenance of equipment, which was outside the framework of this book. In particular, if redundant elements are not placed in service by simple switching, but by physical replacement of the failed components, the problem of anticipating the number of redundant elements is nothing but that of assigning the number of replacement parts to be stocked for maintenance of the system. Thus we shall not begin here a study of passive redundance.

32 Active Redundance at the Level of Substructures or at the Level of Components

Consider a substructure with n components; two simple methods of redundance may be considered in order to form a system that is more reliable than the substructure:

(1) arrange k identical substructures in parallel;

(2) arrange k components in parallel instead of each component of the substructure.

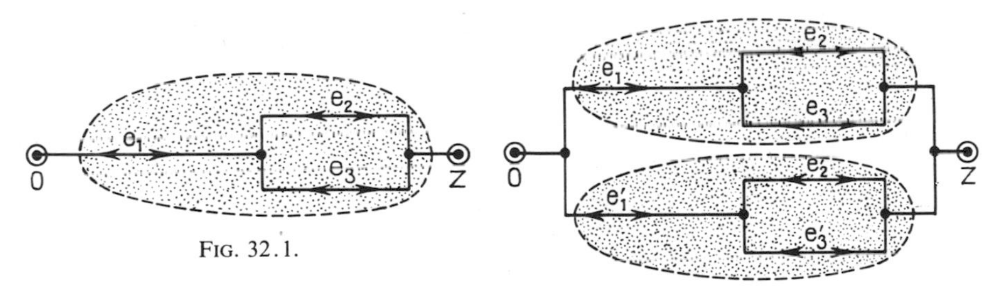

FIG. 32.1.

FIG. 32.2.

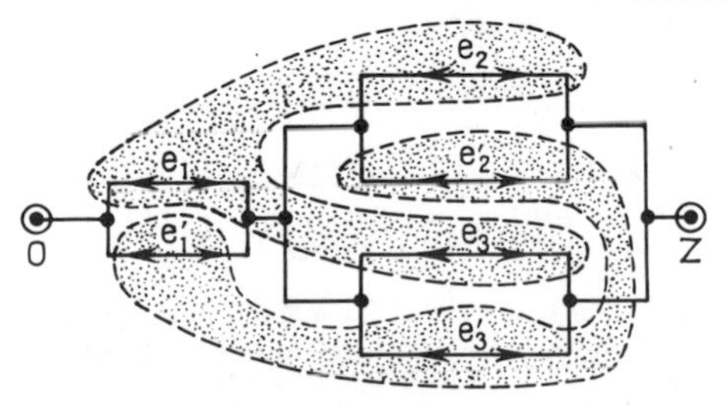

FIG. 32.3.

Example. Consider the substructure whose reliability network is given in Fig. 32.1. In Fig. 32.2 we show the network obtained by putting an identical network in parallel with the given network. On the other hand, in Fig. 32.3 each component e_i, $i = 1, 2, 3$, is duplicated by an identical component e'_i, $i = 1, 2, 3$, in parallel.

We now show that redundance is more effective at the component level than at the substructure level if the base substructure is monotone.

Theorem 32.I. *Let $\varphi(x_1, x_2, \ldots, x_n)$ be a structure function with reliability function $h(p_1, p_2, \ldots, p_n)$ associated with it. Let:*

S be the system obtained by placing in parallel k identical modules[1] *$\varphi(x_1^{(j)}, \ldots, x_n^{(j)})$, $j = 1, \ldots, k$, with the reliabilities of the homologous components $e_i^{(1)}, \ldots, e_i^{(k)}$ being equal,*

$$p_i^{(j)} = p_i, \qquad j = 1, \ldots, k\,; \tag{32.1}$$

C be the system obtained by replacing each component e_i of the structure φ by k distinct components in parallel $e_i^{(1)}, \ldots, e_i^{(k)}$, with the same reliability p_i (cf. (32.1)).

Then, if the structure φ is monotone, the systems S and C have monotone structures and the reliability of C is greater than or equal to that of S.

The system S is obtained by composition (see Section 26) of k modules identical to φ in a parallel structure:

$$\gamma(x^{(1)}, \ldots, x^{(k)}) = 1 - \prod_{j=1}^{k} (1 - x^{(j)})\,. \tag{32.2}$$

Its structure function is therefore

(32.3)

$$\varphi_S(x_1^{(1)}, \ldots, x_n^{(1)}, x_1^{(2)}, \ldots, x_n^{(k)}) = \gamma[\varphi(x_1^{(1)}, \ldots, x_n^{(1)}), \ldots, \varphi(x_1^{(k)}, \ldots, x_n^{(k)})]\,.$$

The system C is obtained by composing n modules identical to γ in the structure φ. Its structure function is therefore

(32.4)

$$\varphi_C(x_1^{(1)}, \ldots, x_1^{(k)}, x_2^{(1)}, \ldots, x_n^{(k)}) = \varphi[\gamma(x_1^{(1)}, \ldots, x_1^{(k)}), \ldots, \gamma(x_n^{(1)}, \ldots, x_n^{(k)})]\,.$$

If the structure φ is monotone, it is representable by a reliability network; it is then the same for φ_S and φ_C according to Theorem 26.I. On the other

[1] Recall (see Section 26) that in the composition of structures we use the term "module" whenever the composed substructures do not have any common component.

hand, let $h(p_1, \ldots, p_n)$ be the reliability function associated with the structure φ, and

$$l(p_i) = 1 - (1 - p_i)^k \tag{32.5}$$

the reliability function associated with the structure γ. Whenever all the homologous components have the same reliability p_i, the reliability functions of the systems S and C are, respectively, $l[h(p_1, \ldots, p_n)]$ and $h[l(p_1), \ldots, l(p_n)]$, and the theorem implies that

$$l[h(p_1, \ldots, p_n)] \leqslant h[l(p_1), \ldots, l(p_n)] . \tag{32.6}$$

For the proof, we show that

$$\varphi_S \leqslant \varphi_C . \tag{32.7}$$

Since the inequality is obvious for any state of the set of components such that $\varphi_S = 0$, we shall consider a link of φ_S. The structure γ being parallel, this link includes a link of at least one of the modules $\varphi^{(j)}$, say $\{ e_{i_1}^{(j)}, e_{i_2}^{(j)}, \ldots, e_{i_s}^{(j)} \}$. Then, however, the modules $\gamma_{i_1}, \gamma_{i_2}, \ldots, \gamma_{i_s}$ of the structure φ_C function since they have a parallel structure and each of them has at least one component in a good state; since they constitute by hypothesis a link of φ, one has $\varphi_C = 1$. In other words, any link of φ_S is a link of φ_C, which proves (32.7). One also sees that the advantage of redundance at the component level arises from the fact that it introduces not only all the "homogeneous" minimal links (that is, those formed of components having the same superior index j) that occur in the structure φ_S, but also "heterogeneous" links obtained by choosing arbitrarily the superior index of the components.

Remarks.

(1) One may also prove Theorem 32.I by recurrence on the number of components using linear composition (Section 26, p. 129).

(2) If φ is not degenerate, if $k \geqslant 2$, and if

$$(p_1, \ldots, p_n) \neq (0, \ldots, 0) \tag{32.8}$$

and

$$(p_1, \ldots, p_n) \neq (1, \ldots, 1) , \tag{32.9}$$

inequality (32.6) holds strictly.

(3) One may interpret Theorem 32.I in terms of "cannibalization" of systems. If one has at one's disposal k identical pieces of equipment, of which two have failed because of failure of different components, repairs may be made using some good components from one in the other. More generally, *cannibalization* consists in transferring some components of one system into another in such a way as to maintain in a functioning state the

largest possible number of pieces of equipment (see references [31, 51, 52]). The interest in cannibalization derives from the property expressed in Theorem 32.I, this operation amounting to carrying out a redundance at the level of components on an ad hoc basis.

Example. We seek to determine the reliability of a radar indicator I, the probability p that it functions 100 hr without failure. The indicator I may be subdivided into five components (drawers and chassis), all necessary for functioning (a structure having a reliability network formed of five elements in series). The reliability of each of these components is given in Table 32.I.

Component number	Reliability (probability of functioning 100 hr without failure)
1	$p_1 = 0.96$
2	$p_2 = 0.93$
3	$p_3 = 0.85$
4	$p_4 = 0.80$
5	$p_5 = 0.75$
Indicator I	$\pi = 0.455$

TABLE 32.I

If we duplicate the indicator (two identical indicators placed side by side), the reliability of the set is

$$(32.10) \qquad h_1(p) = 1 - (1 - \pi)^2 = 1 - (1 - 0.455)^2 = 0.703\,.$$

If we duplicate each component independently, we obtain

$$(32.11) \qquad h_2(p) = \prod_{i=1}^{5} [1-(1-p_i)^2]$$
$$= (1-0.04^2)\,(1-0.07^2)\,(1-0.15^2)\,(1-0.20^2)\,(1-0.25^2)$$
$$= 0.874\,.$$

Redundance of the components is clearly more effective than redundance of the indicator I.

33 Optimal Redundance

When designing a system with a view toward a given use, one may consider the problem of optimal redundance, that is, of the choice of the number of redundant components to be put in place to render maximal or minimal a certain function that corresponds to a particular criterion. Such criteria might be:

(1) the reliability of the system itself, this for a given total number of components;
(2) the total cost for a fixed reliability, with a constraint on the number of components.

Or some other criteria on weight, size, etc., might apply.

Ever since Pareto we know that the optimization of several economic functions at the same time, from a mathematical point of view, may not have sense. Generally, one is given one and only one economic function and a set of constraints, and the problem consists of determining the optimum (maximum or minimum, depending on the case) of the economic function with respect to the constraints. The optimal solution may or may not be unique depending on the nature of the problem. In certain cases one can extend these optimization problems to those called "optimation with parameters," that is, to the study of the evolution of the optimum by varying a parameter, or several. A great diversity of optimization problems may be posed using the notion of redundant systems, and we shall limit ourselves here to a few of them. In particular, in that which follows we shall consider as functions to be optimized the reliability of systems, or the number of components and some constraints or specifications on the costs; but instead of costs one can also consider the weight or size, or any other physical or technical aspect.

The search for the optimal redundance will be guided by the result of Theorem 32.I, where it was shown that one should always seek an improvement at the component level and not at the levels of the complete structure or of substructures.

Search for the Minimal Number of Redundant Components Starting from a Series Structure. We intend to determine this minimal number for a redundant structure having a reliability greater than or equal to a given reliability, obtained from a series structure.

We proceed using a sequential method proposed by Barlow and Proschan [5]. This method may be applied to a basic structure (φ, in the notation of Theorem 32.I) other than a series structure, but then it does not necessarily lead to an optimum.

Let S be the initial series structure, and denote by S_i the new structure obtained by doubling (that is, by placing an identical component in parallel) component e_i.

We shall choose the component e_i that maximizes the reliability of S_i, then we shall denote by S_{ij} the structure obtained by doubling component

e_j after having doubled component e_i, and we shall choose e_j so that the reliability of S_{ij} is maximal and continue with this until the desired reliability has been attained. For a series structure, this procedure leads to the optimum in a rigorous fashion (a proof will be given later in the study of redundance with minimal cost).

Suppose that the system S is composed of n components $e_1, e_2, \ldots, e_n$ in series and that the respective reliabilities are $p_1, p_2, \ldots, p_n$. The reliability of the system S is then

$$h(p_1, p_2, \ldots, p_n) = p_1 \cdot p_2 \cdot \ldots \cdot p_n . \tag{33.1}$$

If we double the component e_i, the reliability of the new system S_i is then

$$\begin{aligned} h_i(p_1, p_2, \ldots, p_i, \ldots, p_n) &= p_1 \cdot p_2 \cdot \ldots \cdot [1 - (1 - p_i)^2] \cdot \ldots \cdot p_n \\ &= p_1 \cdot p_2 \cdot \ldots \cdot p_i(2 - p_i) \cdot \ldots \cdot p_n \\ &= (2 - p_i) \cdot p_1 \cdot p_2 \cdot \ldots \cdot p_i \cdot \ldots \cdot p_n \\ &= (2 - p_i) \cdot h(p_1, p_2, \ldots, p_i, \ldots, p_n) . \end{aligned} \tag{33.2}$$

Thus $h_i(p_1, p_2, \ldots, p_n)$ is greater the smaller p_i is. Therefore we want to double the weakest component. Repeating this reasoning, we see that at the second stage, we should either add another component e_i or double the least reliable component other than e_i, and so on.

Examples.

(1) Consider the network of Fig. 33.1, for which the total reliability has the value 0.34. We intend to add components e_1, e_2, e_3 in minimal number so that the reliability becomes at least 0.70. The components will be added only in parallel to the components already existing and will be identical to them.

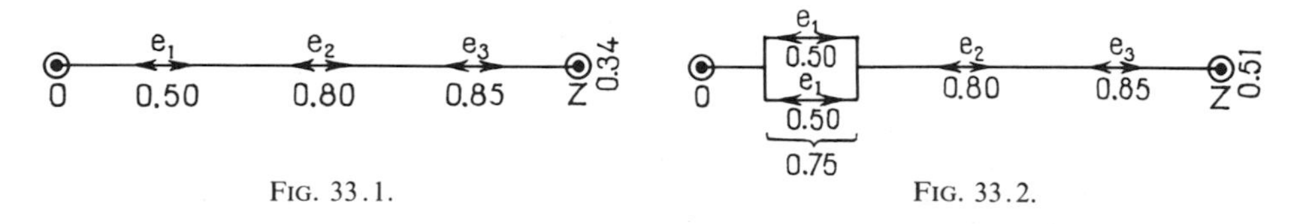

FIG. 33.1. FIG. 33.2.

According to the rule given above, we double element e_1 whose reliability is the weakest. We then obtain the network of Fig. 33.2, for which we have a total reliability of 0.51.

To continue, it is necessary to compare the two results: triple e_1 (Fig. 33.3) or double e_2 (Fig. 33.4). It is useless to examine the result of doubling e_3 since the reliability of e_2 is less than that of e_3. For Fig. 33.3 we have 0.595, and for Fig. 33.4 we have 0.61. We continue therefore from the latter.

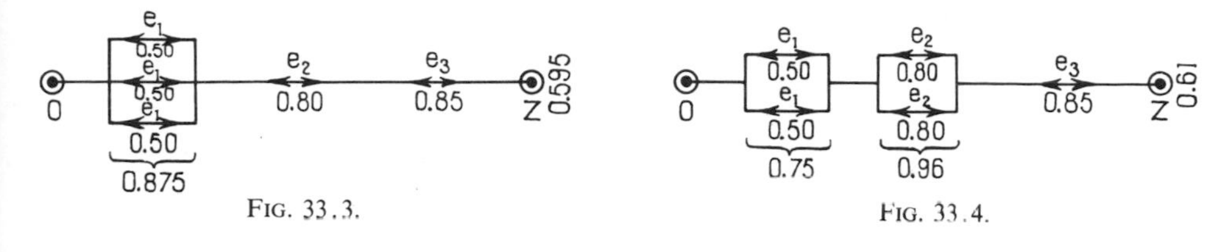

FIG. 33.3.

FIG. 33.4.

We now examine which is better:

(a) triple e_1 (Fig. 33.5), or
(b) double e_3 (Fig. 33.6).

The corresponding results are 0.713 and 0.703. The results of Figs. 33.5 and 33.6 both suffice since the reliabilities obtained exceed 0.70; but, to choose, the network of Fig. 33.5 is better since there we have the greater reliability with three redundant elements.

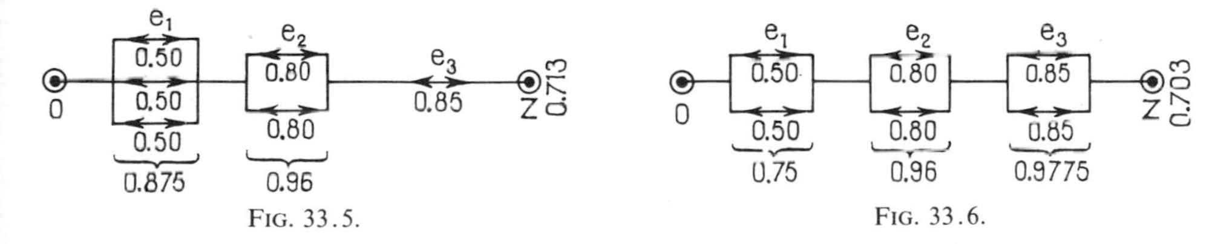

FIG. 33.5.

FIG. 33.6.

(2) Next we consider an example of Section 32 (radar indicator). We intend to exceed a reliability of 0.98. In Table 33.I we present the optimal number of redundant components and the reliabilities obtained. The components are arranged from left to right in the order of decreasing reliabilities; the number of redundant components is therefore nondecreasing from left to right.

One sees that the best solution is given by 10′, that is, by the following redundances.

e_1: doubled,
e_2: doubled,
e_3: tripled,
e_4: quadrupled,
e_5: quadrupled.

Note. If at the nth stage of the calculations we have obtained an optimal state with k_i redundant components e_i and k_j with $k_i \leqslant k_j$, and if the reliability of k_i components e_i in parallel is less than the reliability of k_j components e_j, then at state $n+1$ we would prefer to add an e_i component rather than an e_j component.

This remark lets us economize significantly on the number of stages in

TABLE 33.I. Number of Redundant Components and Reliability of the Module

Stage	e_1	e_2	e_3	e_4	e_5	Reliability of the system
0	0 0.96	0 0.93	0 0.85	0 0.80	0 0.75	0.455
1	0 0.96	0 0.93	0 0.85	0 0.80	1 0.937	0.569
2	0 0.96	0 0.93	0 0.85	0 0.80	2 0.984	0.598
2′ (optimal)	0 0.96	0 0.93	0 0.85	1 0.96	1 0.937	0.683
3	0 0.96	0 0.93	0 0.85	1 0.96	2 0.984	0.717
3′ [a]	0 0.96	0 0.93	0 0.85	2 0.992	1 0.937	0.705
3″ (optimal)	0 0.96	0 0.93	1 0.977	1 0.96	1 0.937	0.785
4	0 0.96	0 0.93	1 0.977	1 0.96	2 0.984	0.825
4′ (optimal)	0 0.96	1 0.995	1 0.977	1 0.96	1 0.937	0.840
5 (optimal)	0 0.96	1 0.995	1 0.977	1 0.96	2 0.984	0.882
5′	1 0.998	1 0.995	1 0.977	1 0.96	1 0.937	0.874
6 (optimal)	1 0.998	1 0.995	1 0.977	1 0.96	2 0.984	0.918
7	1 0.998	1 0.995	1 0.977	1 0.96	3 0.996	0.929
7′ (optimal)	1 0.998	1 0.995	1 0.977	2 0.992	2 0.984	0.948
8	1 0.998	1 0.995	1 0.977	2 0.992	3 0.996	0.959
8′ (optimal)	1 0.998	1 0.995	2 0.998	2 0.992	2 0.984	0.967
9 (optimal)	1 0.998	1 0.995	2 0.998	2 0.992	3 0.996	0.978
9′	1 0.998	2 0.999	2 0.998	2 0.992	2 0.984	0.971
10	1 0.998	1 0.995	2 0.998	2 0.992	4 0.999	0.981
10′ (optimal)	1 0.998	1 0.995	2 0.998	3 0.998	3 0.996	0.985
10″	1 0.998	2 0.999	2 0.998	2 0.992	3 0.996	0.983

[a] This stage is useless since $p_5 < p_4$; it is certain that 3′ will be at least as good as 3. We have presented this stage in the interest of pedagogy.

the calculation. In the example above, one may go directly from stage 1 to 4′, then from 5 to 8′, and finally from 9′ to 10′. We thus suppress 12 of the 21 stages of the calculation.

Redundance at Minimal Cost. We now propose to determine the total minimal cost of redundant components, requiring that the reliability of the system be greater than or equal to a given reliability. Note that in the case where the unit cost of all the components is equal to 1, we return to the preceding problem, which is thus a particular case.

We shall represent by the n-tuple

$$k = (k_1, k_2, \dots, k_n) \tag{33.3}$$

the system obtained by placing k_i components in parallel with the component e_i of the initial structure ($i = 1, \dots, n$), and designate by $r(k)$ the reliability of this system.

As before, the algorithm will consist of progressively adding components until the desired reliability has been obtained. If, at an arbitrary stage, k represents the number of components of each type that have already been added, the index i of the supplementary component will be chosen in such a way as to maximize the quantity

$$\mu_i = \frac{1}{c_i}\,[\log r(k_1, \dots, k_i + 1, \dots, k_n) - \log r(k)] \tag{33.4}$$

where c_i is the unit cost of component e_i.

We go on to see that this algorithm leads to the optimal redundance in the case where the basic structure one starts with is a series structure with reliability

$$r(0, \dots, 0) = h(p_1, \dots, p_n) = p_1 \cdot p_2 \cdot \dots \cdot p_n \tag{33.5}$$

where p_i is the reliability of component e_i. In this case the reliability of the system obtained by redundance k is

$$r(k) = \prod_{i=1}^{n} r_i(k_i)\,, \tag{33.6}$$

with

$$r_i(k_i) = 1 - (1 - p_i)^{k_i + 1}\,. \tag{33.7}$$

We may easily show, however, expression (33.7) is a concave function of k_i (the proof will be given in a more general framework in Section 34); this then holds also for $\log r_i(k_i)$, that is,

$$\log r_i(k_i + 1) - \log r_i(k_i) \leqslant \log r_i(k_i) - \log r_i(k_i - 1)\,. \tag{33.8}$$

It then follows that the algorithm described above, which consists of adding at each stage a component e_i such that

(33.9)

$$\mu_i = \frac{1}{c_i}\left[\log r(k_1, \ldots, k_i + 1, \ldots, k_n) - \log r(k)\right] = \frac{1}{c_i}\left[\log r_i(k_i + 1) - \log r_i(k_i)\right]$$

is maximal, amounts to classifying in decreasing order the set of quantities

$$\mu_i(k_i) = \frac{1}{c_i}\left[\log r_i(k_i + 1) - \log r_i(k_i)\right] \tag{33.10}$$

for all values of i ($i = 1, \ldots, n$) and all values of k_i ($k_i = 0, 1, 2, \ldots$), and adding at each stage the component that has furnished the following term in the list (see the example given later). If e_s is the last component added, whose addition has for the first time caused the reliability to exceed the given one, and if the redundance thus obtained is

$$k^* = (k_1^*, \ldots, k_n^*)\,, \tag{33.11}$$

then we have

$$\mu_s(k_s^* - 1) \geqslant \mu_i(k_i^*)\,, \qquad i = 1, \ldots, n\,, \tag{33.12}$$

that is,

(33.13)

$$\frac{1}{c_s}\left[\log r_s(k_s^*) - \log r_s(k_s^* - 1)\right] \geqslant \frac{1}{c_i}\left[\log r_i(k_i^* + 1) - \log r_i(k_i^*)\right].$$

Consider now a redundance k different than k^*; we proceed to show that one cannot have simultaneously

$$r(k) > r(k^*) \quad \text{and} \quad \sum_{i=1}^{n} k_i c_i \leqslant \sum_{i=1}^{n} k_i^* c_i\,. \tag{33.14}$$

Let $\boldsymbol{I}_1$ be the set of indices i such that $k_i > k_i^*$, and $\boldsymbol{I}_2$ the set of indices i such that $k_i < k_i^*$. We then have, using the properties of concavity of $r_i(k_i)$,

(33.15)

$$\begin{aligned}
\log r(k) - \log r(k^*) &= \sum_{i \in \boldsymbol{I}_1}\left[\log r_i(k_i) - \log r_i(k_i^*)\right] \\
&\qquad - \sum_{i \in \boldsymbol{I}_2}\left[\log r_i(k_i^*) - \log r_i(k_i)\right] \\
&\leqslant \sum_{i \in \boldsymbol{I}_1}(k_i - k_i^*)\left[\log r_i(k_i^* + 1) - \log r_i(k_i^*)\right] \\
&\qquad - \sum_{i \in \boldsymbol{I}_2}(k_i^* - k_i)\left[\log r_i(k_i^*) - \log r_i(k_i^* - 1)\right] \\
&\leqslant \sum_{i=1}^{n}(k_i - k_i^*)\left[\log r_i(k_i^* + 1) - \log r_i(k_i^*)\right]
\end{aligned}$$

or

$$\log r(k) - \log r(k^*) \leqslant \sum_{i=1}^{n} (k_i - k_i^*)\, c_i\, \mu_i(k_i^*) \tag{33.16}$$

and finally, using (33.12),

$$\log r(k) - \log r(k^*) \leqslant \sum_{i=1}^{n} (k_i - k_i^*)\, c_i\, \mu_s(k_s^* - 1)\,. \tag{33.17}$$

It then follows that

$$r(k) > r(k^*) \Rightarrow \sum_{i=1}^{n} k_i\, c_i > \sum_{i=1}^{n} k_i^*\, c_i\,, \tag{33.18}$$

which proves the impossibility of (33.14).

It is therefore not possible to obtain a reliability greater than with the redundance k^* without additional expense. Note, however, that there may exist a redundance k such that $r(k)$ is less than $r(k^*)$, but greater than the expected reliability, and for which the cost is less than that of k^* (see [22] for more on this subject).

Example. We consider again Example 2 on p. 175, now taking into account the costs c_i of the redundant elements. Table 33.II indicates the stages of the calculation when one applies the algorithm as originally described and supposing that the required reliability is 0.80. For each stage, we have indicated the value of the quantities $\mu_i(k_i)$ where k_i is the number of redundant components e_i added in the preceding stages; then the new redundance k is

Component :		e_1	e_2	e_3	e_4	e_5	Reliability of the system	Cost of the redundance
Reliability		0.96	0.93	0.85	0.8	0.75	0.455	0
Cost C_i		300	1 200	800	500	1 000		
First stage	$\mu_i . 10^5$	5.62	2.45	7.56	15.8*	9.67		
	k_i^*	0	0	0	1	0		500
	$r_i(k_i^*)$	0.96	0.93	0.85	0.96	0.75	0.546	
Second stage	$\mu_i . 10^5$	5.62	2.45	7.56	2.85	9.67*		
	k_i^*	0	0	0	1	1		1 500
	$r_i(k_i^*)$	0.96	0.93	0.85	0.96	0.937	0.683	
Third stage	$\mu_i . 10^5$	5.62	2.45	7.56*	2.85	2.13		
	k_i^*	0	0	1	1	1		2 300
	$r_i(k_i^*)$	0.96	0.93	0.977	0.96	0.937	0.785	
Fourth stage	$\mu_i . 10^5$	5.62*	2.45	1.2	2.85	2.13		
	k_i^*	1	0	1	1	1		2 600
	$r_i(k_{i^*})$	0.998	0.93	0.977	0.96	0.937	0.816	

TABLE 33.II

given, obtained by adding the component for which μ_i is maximal, and then the reliability is obtained.

As we have remarked above, since at each stage one chooses the component i to add according to the largest $\mu_i(k_i)$ not yet used, it is more convenient to set up for each component i a table of values $\mu_i(k_i)$, then collect these tables in a unique list in decreasing order, as in Table 33.III. The order in which one should add components is then simply that indicated in the first column. Figure 33.7 gives the curve of the maximal reliability as a function of price (or of minimal cost as a function of reliability), traced from the successively obtained solutions.

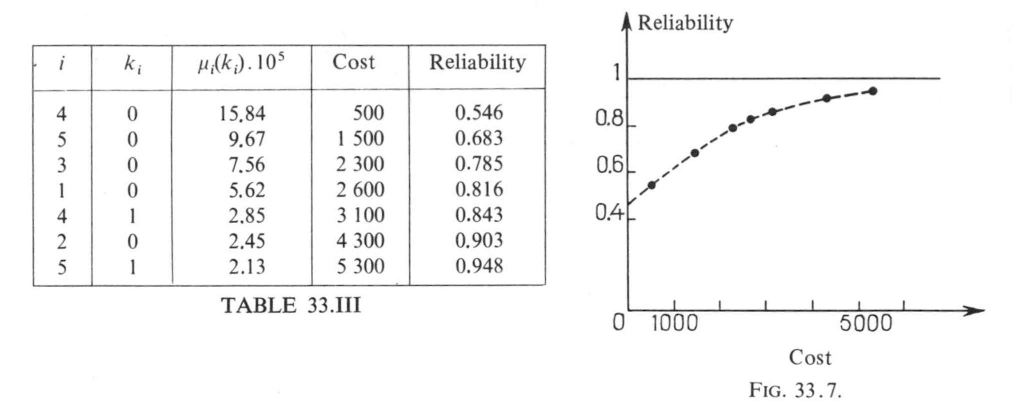

i	k_i	$\mu_i(k_i).10^5$	Cost	Reliability
4	0	15.84	500	0.546
5	0	9.67	1 500	0.683
3	0	7.56	2 300	0.785
1	0	5.62	2 600	0.816
4	1	2.85	3 100	0.843
2	0	2.45	4 300	0.903
5	1	2.13	5 300	0.948

TABLE 33.III

FIG. 33.7.

34 Concavity of Monotone Structures with Respect to Redundance

Let $h(p_1, p_2, \ldots, p_n)$ be the reliability function of a monotone structure $\varphi(x_1, x_2, \ldots, x_n)$ and $r(k_1, k_2, \ldots, k_n; p_1, p_2, \ldots, p_n)$ the reliability function of the structure

(34.1)

$$\varphi_k(x_1^{(0)}, \ldots, x_1^{(k_1)}, x_2^{(0)}, \ldots, x_n^{(k_n)}) = \varphi[\gamma(x_1^{(0)}, \ldots, x_1^{(k_1)}), \ldots, \gamma(x_n^{(0)}, \ldots, x_n^{(k_n)})]$$

where γ is the parallel structure (32.2). The structure φ_k, where

(34.2) $$k = (k_1, \ldots, k_n),$$

is derived from φ by adding k_i identical components in parallel with e_i ($k_i \geqslant 0$).

Theorem 34.I. *The function $r(k_1, \ldots, k_n; p_1, \ldots, p_n)$ is concave with respect to each of the variables k_i, that is,*

(34.3)

$$r(k_1, \ldots, k_i, \ldots, k_n; p_1, \ldots, p_n) - r(k_1, \ldots, k_i - 1, \ldots, k_n; p_1, \ldots, p_n)$$
$$\geqslant r(k_1, \ldots, k_i + 1, \ldots, k_n; p_1, \ldots, p_n) - r(k_1, \ldots, k_i, \ldots, k_n; p_1, \ldots, p_n)\,.$$

Proof. Note first that, as in the particular case of Theorem 32.I (where we had $k_1 = k_2 = \cdots = k_n$), the fact that φ is monotone, therefore representable by a reliability network, implies that φ_k is likewise representable by a network, thus monotone. Moreover, we have seen in Section 26 (p. 129), discussing monotone linear composition, that one may single out an arbitrary component $e_i^{(j)}$ of φ_k; in fact, to recover relation (34.3) we shall consider the structure $\varphi_{k'}$ with

$$k' = (k_1, \ldots, k_i - 1, \ldots, k_n) . \tag{34.4}$$

We obviously suppose that $k_i \geqslant 1$. Let $e_i^{(j)}$ be one of the k_i components identical to e_i of the structure φ_k; the reliability network of this structure may then be put in the form indicated in Fig. 34.1, analogous to Fig. 26.11. This corresponds, for the reliability functions, to the relation

$$r_{k_i - 1} = p_i f + (1 - p_i) g \tag{34.5}$$

where

$$r_{k_i - 1} = r(k_1, \ldots, k_i - 1, \ldots, k_n ; p_1, \ldots, p_n) \tag{34.6}$$

and where f and g are reliability functions of the structures represented by the networks $\mathfrak{R}_1$ and $\mathfrak{R}_2$. These functions, according to Theorem 26.V, satisfy the inequality

$$f \geqslant g . \tag{34.7}$$

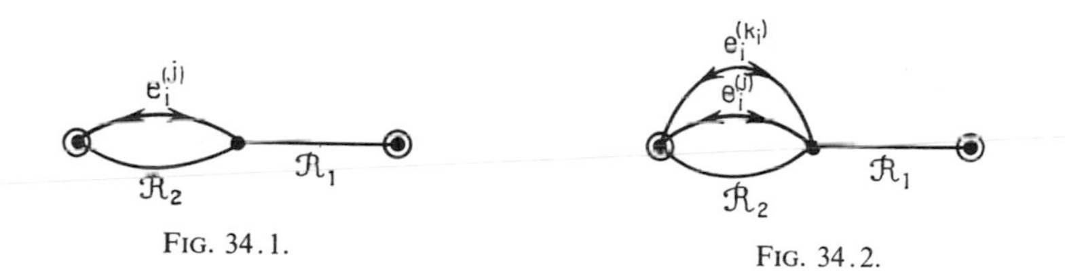

Fig. 34.1.

Fig. 34.2.

The structure φ_k is obtained by adding in parallel to the components $e_i^{(0)}, e_i^{(2)}, \ldots, e_i^{(k_i - 1)}$ a supplementary component $e_i^{k_i}$ (Fig. 34.2); its reliability function is therefore

$$r_{k_i} = [1 - (1 - p_i)^2] f + (1 - p_i)^2 g . \tag{34.8}$$

By adding still another component identical to e_i, we obtain

$$r_{k_i + 1} = [1 - (1 - p_i)^3] f + (1 - p_i)^3 g , \tag{34.9}$$

from which

$$r_{k_i} - r_{k_i - 1} = p_i(1 - p_i)(f - g), \tag{34.10}$$

$$r_{k_i + 1} - r_{k_i} = p_i(1 - p_i)^2 (f - g) = (1 - p_i)(r_{k_i} - r_{k_i - 1}) . \tag{34.11}$$

Since $0 \leqslant p_i \leqslant 1$ and $f - g \geqslant 0$, we see that:

(a) $r_{k_i} - r_{k_i - 1} \geqslant 0$: active redundance always improves the reliability of a monotone structure;

(b) $r_{k_i} - r_{k_i - 1} \geqslant r_{k_i + 1} - r_{k_i}$: the reliability function is concave; that is, the increase in reliability is smaller and smaller as one adds components. Redundance thus has decreasing "yield."

Example. As an example we take one of the monotone structures with three components presented in Fig. 26.13 (p. 134); specifically that one denoted $\varphi_2^{(3)}$:

$$\varphi(x_1, x_2, x_3) = x_3 + x_1 x_2 - x_1 x_2 x_3 . \tag{34.12}$$

The reliability network is represented in Fig. 34.3. The proof of Theorem 34.I leads to putting the structure function in the form

$$\begin{aligned} \varphi(x_1, x_2, x_3) &= x_1\, \varphi(1, x_2, x_3) + (1 - x_1)\, \varphi(0, x_2, x_3) \\ &= x_1(x_2 + x_3 - x_2 x_3) + (1 - x_1)\, x_3 , \end{aligned} \tag{34.13}$$

if x_1 is the component that one singles out. The expression $x_2 + x_3 - x_2 x_3$ corresponds to placing components e_2 and e_3 in parallel; the reliability network corresponding to (34.13) is that of Fig. 34.4, where we see the appearance of the subnetworks $\mathfrak{R}_1$ and $\mathfrak{R}_2$ of Fig. 34.1. One may easily verify that the networks of Figs. 34.3 and 34.4 are equivalent.

FIG. 34.3. FIG. 34.4.

We return now to expression (34.12). The reliability function of the system is

$$\begin{aligned} h(p_1, p_2, p_3) &= p_3 + p_1 p_2 - p_1 p_2 p_3 \\ &= p_3 + p_1 p_2(1 - p_3) . \end{aligned} \tag{34.14}$$

Suppose, for example, that we have

$$p_1 = 0.4, \qquad p_2 = 0.7, \qquad p_3 = 0.5 . \tag{34.15}$$

It then follows that

$$h = 0.5 + 0.35\, p_1, \tag{34.16}$$

that is,

$$h = 0.64 . \tag{34.17}$$

If component e_1 is doubled, the reliability obtained is

$$r(1, 0, 0) = 0.5 + 0.35[1 - (1 - p_1)^2] = 0.724 ,$$

an increment of $0.724 - 0.640 = 0.084$.

Adding another component e_1 in parallel with the first two gives

$$r(2, 0, 0) = 0.5 + 0.35[1 - (1 - p_1)^3] = 0.774\,4 ,$$

which is a new increment of $0.7744 - 0.7240 = 0.0504$, less than the preceding one. The reader may verify these results using (34.10) and (34.11) where

$$f = p_2 + p_3 - p_2 p_3 = 0.85 ,$$
$$g = p_3 = 0.5 ,$$
$$p_i = p_1 = 0.4 .$$

35 Type k of n Structures

Definition. *We shall say that a structure function* $\varphi(x_1, x_2, \dots, x_n)$ *is of type "k of n" and denote this as* $\varphi_n^k(x)$ *if*

$$(S(x) \geqslant k) \Rightarrow \varphi_n^k(x) = 1 , \tag{35.1}$$

$$(S(x) < k) \Rightarrow \varphi_n^k(x) = 0 ; \tag{35.2}$$

where

$$1 \leqslant k \leqslant n \tag{35.3}$$

and

$$S(x) = x_1 + x_2 + \cdots + x_n . \tag{35.4}$$

Example. We show that the structure function

$$\varphi(x_1, x_2, x_3) = x_1 x_2 + x_2 x_3 + x_3 x_1 - 2 x_1 x_2 x_3 \tag{35.5}$$

is a function of type 2 of 3. For this, we establish its table of values (Fig. 35.1). Note that for $S(x) \geqslant 2$ we have $\varphi(x) = 1$, and that for $S(x) < 2$ we have $\varphi(x) = 0$.

Now we consider a few obvious properties of type k of n structures.

x_1	x_2	x_3	$x_1\,x_2$	$x_2\,x_3$	$x_3\,x_1$	$2\,x_1\,x_2\,x_3$	$S(x)$	$\varphi(x)$
0	0	0	0	0	0	0	0	0
0	0	1	0	0	0	0	1	0
0	1	0	0	0	0	0	1	0
0	1	1	0	1	0	0	2	1
1	0	0	0	0	0	0	1	0
1	0	1	0	0	1	0	2	1
1	1	0	1	0	0	0	2	1
1	1	1	1	1	1	2	3	1

FIG. 35.1.

(1) A system whose structure function is $\varphi_n^k(x)$ functions if the number of components in good states is at least equal to k. In particular

if $k = 1$, one has a parallel structure;
if $k = n$, one has a series structure.

(2) $\varphi_n^k(x)$ is monotone (the structure function of the example above is included among the structure functions of order 3 enumerated in Fig. 26.13, p. 134).

(3) Any subset of at least k components is a link of $\varphi_n^k(x)$. Any subset of k components is a minimal link.

(4) There are $\binom{n}{i}$ links with k components ($i \geqslant k$).

(5) There are $\binom{n}{k}$ minimal links having k components.

(6) The reliability network $\mathfrak{R}_n^k$ corresponding to a structure function $\varphi_n^k(x)$ is obtained by arranging in parallel the $\binom{n}{k}$ minimal links $\mathbf{a}_1, \mathbf{a}_2, \ldots, \mathbf{a}_{\binom{n}{k}}$ each made up of k components (Fig. 35.2).

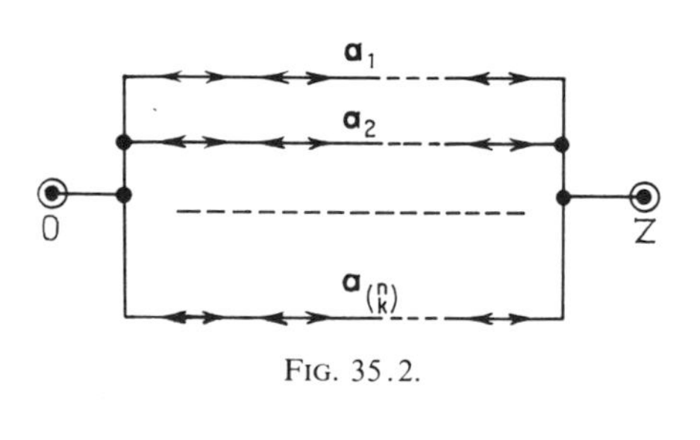

FIG. 35.2.

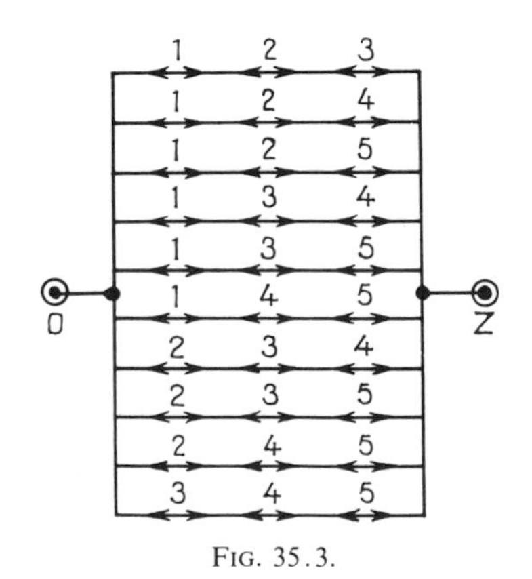

FIG. 35.3.

(7) The structure function $\varphi_n^k(x)$ may be written (cf. (22.1)) as

$$\varphi_n^k(x) = 1 - \prod(1 - x_{a_1} x_{a_2} \dots x_{a_k}) \tag{35.6}$$

where the product is extended to the $\binom{n}{k}$ minimal links, that is, to the $\binom{n}{k}$ combinations of k components.

(8) $\varphi_n^k(x)$ is a symmetric function of the variables $x_1, x_2, \dots, x_n$.

Example. We consider an example that illustrates the above properties. Let $\mathfrak{R}_5^3$ be a system whose components constitute the set $\{ e_1, e_2, e_3, e_4, e_5 \}$. The network is made up of $\binom{5}{3} = 10$ links in parallel (Fig. 35.3). We note that any subset with three components or more is a link.

There are $\binom{5}{3} = 10$ minimal links, $\binom{5}{4}$ links with four components, and $\binom{5}{5} = 1$ link with five components. The structure function is

(35.7)

$$\begin{aligned}\varphi_n^k(x_1, x_2, x_3, x_4, x_5) &= 1 - (1 - x_1 x_2 x_3).(1 - x_1 x_2 x_4)\\ &\times(1 - x_1 x_2 x_5).(1 - x_1 x_3 x_4).(1 - x_1 x_3 x_5).(1 - x_1 x_4 x_5)\\ &\times(1 - x_2 x_3 x_4).(1 - x_2 x_3 x_5).(1 - x_2 x_4 x_5).(1 - x_3 x_4 x_5)\,.\end{aligned}$$

It is easy to see that this function is symmetric; it is not changed if one permutes an x_i with an x_j, $i \neq j$. For example,

$$\varphi(x_1, x_2, x_3, x_4, x_5) = \varphi(x_1, x_5, x_3, x_4, x_2)\,. \tag{35.8}$$

Reduction of a Type k of n Network. A structure function of type k of n in form (35.6) or the equivalent network following the scheme of Fig. 35.2 is very complicated when n becomes large; it is of interest to be able to simplify these. We therefore seek a network possessing a minimal number of arcs equivalent to a given network $\mathfrak{R}_n^k$.

Hansel [29] has shown that this minimal number of arcs for a network $\mathfrak{R}_n^2$ may be given by

$$E_n^2 = n(m + 2) - 2^{m+1} \tag{35.9}$$

where m is defined by

$$2^m < n \leqslant 2^{m+1}\,. \tag{35.10}$$

Example. Consider the reduced networks $\mathfrak{R}_n^2$, $n = 3, 4, 5$. One obtains

(35.11) $n = 3 \Rightarrow m = 1\,,\quad E_3^2 = 5$ (Fig. 35.4),

(35.12) $n = 4 \Rightarrow m = 1\,,\quad E_4^2 = 8$ (Fig. 35.5),

(35.13) $n = 5 \Rightarrow m = 2\,,\quad E_5^2 = 12$ (Fig. 35.6).

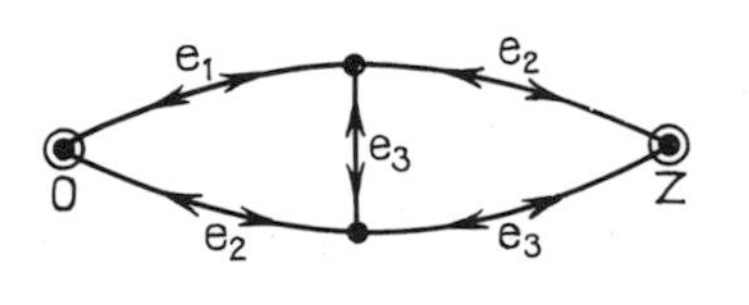

FIG. 35.4.

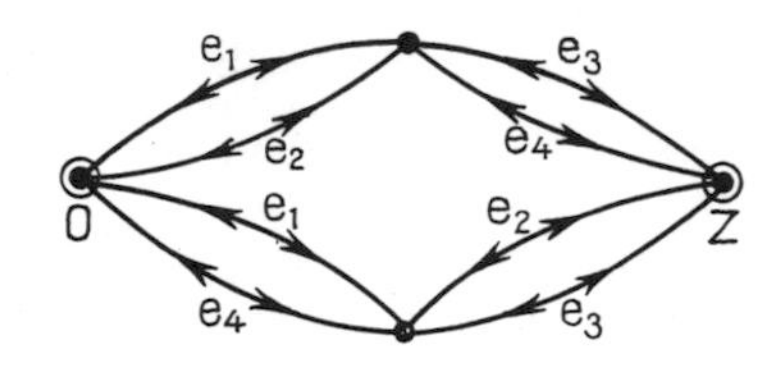

FIG. 35.5.

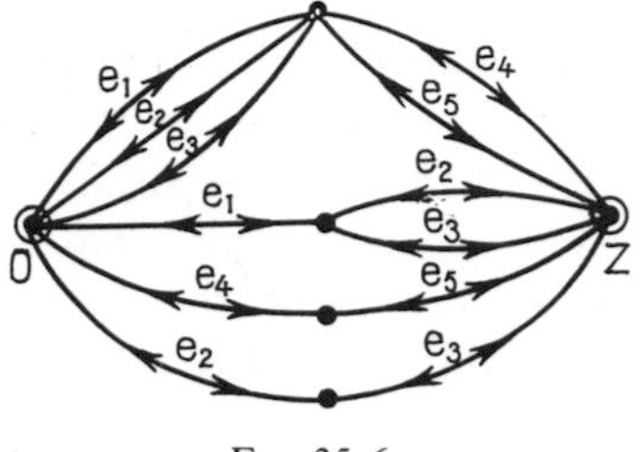

FIG. 35.6.

The following rule permits one to construct easily reduced networks $\mathfrak{R}_n^k$, although not necessarily reduced to a minimum: $\mathfrak{R}_n^k$ is always equivalent to the network represented in Fig. 35.7, where

$$1 \leqslant l \leqslant n - 1, \tag{35.14}$$

$$i = \text{MAX}\,\{\,0, l + k - n\,\}, \tag{35.15}$$

$$j = \text{MIN}\,\{\,k, l\,\}. \tag{35.16}$$

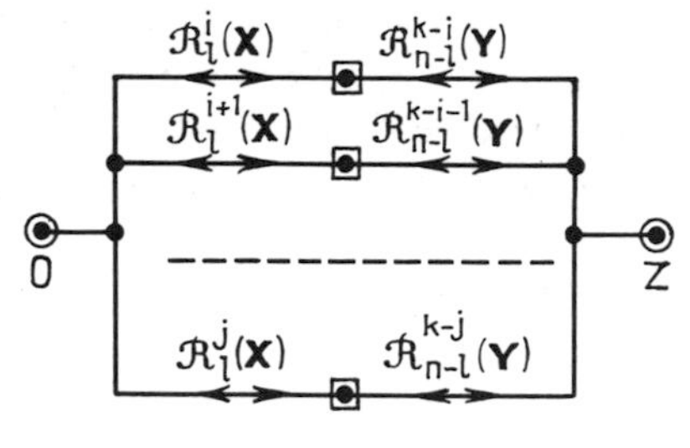

FIG. 35.7.

Here $\mathbf{X} = \{\,1, 2, \ldots, l\,\}$ is the set of the first l components and $\mathbf{Y} = \{\,l + 1, l + 2, \ldots, n\,\}$.

An especially reduced network is obtained by choosing

$$l = \left[\frac{n}{2}\right] \tag{35.17}$$

where $[A]$ is the greatest integer less than A.

Examples. Two explicit examples are given in Figs. 35.8 and 35.9:

(a) 3 of 4 network, for $l = 2$ ($i = 1, j = 2$);
(b) 3 of 5 network, for $l = 2$ ($i = 0, j = 2$).

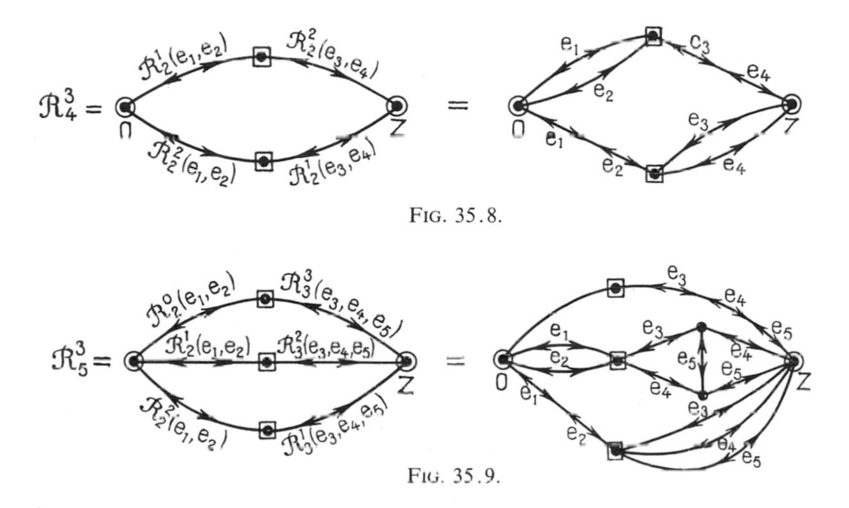

FIG. 35.8.

FIG. 35.9.

Type k of n Structure with Components of the Same Reliability. If the components of a structure $\varphi_n^k(x)$ have the same reliability p, the reliability function may be written, from (25.18) and taking into account property (4) given above,

$$h_n^k(p) = \sum_{i=k}^{n} \binom{n}{i} p^i (1 - p)^{n-i} . \tag{35.18}$$

We go on to see that when $2 \leqslant k \leqslant n - 1$, the curve representative of $h_n^k(p)$ is of type II (p. 138), that is, an S curve; it intersects the first bisector at a point p_0 defined by the equation

$$h_n^k(p_0) = p_0 . \tag{35.19}$$

If $p > p_0$, the type k of n structure is more reliable than a single component; on the contrary, if $p < p_0$, it is less reliable than this single component.

Indeed, we calculate the first and second derivatives of $h_n^k(p)$ expressed by (35.18); after simplification we have

$$\frac{\mathrm{d}}{\mathrm{d}p} h_n^k(p) = \frac{n\,!}{(k-1)\,!\,(n-k)\,!} \cdot p^{k-1} . (1-p)^{n-k} , \tag{35.20}$$

$$\frac{\mathrm{d}^2}{\mathrm{d}p^2} h_n^k(p) = \frac{n\,!}{(k-1)\,!\,(n-k)\,!} \cdot p^{k-2} . (1-p)^{n-k-1} . [k - 1 - (n-1)\,p] . \tag{35.21}$$

According to (35.20) and the condition $2 \leqslant k \leqslant n - 1$, we have $h'(0) = h'(1) = 0$, which shows that the curve is of type II (see (27.19)). On the other hand, we have

$$\left(p < \frac{k-1}{n-1}\right) \Rightarrow \left(\frac{d^2}{dp^2} h_n^k(p) > 0\right), \tag{35.22}$$

$$\left(p = \frac{k-1}{n-1}\right) \Rightarrow \left(\frac{d^2}{dp^2} h_n^k(p) = 0\right), \tag{35.23}$$

$$\left(p > \frac{k-1}{n-1}\right) \Rightarrow \left(\frac{d^2}{dp^2} h_n^k(p) < 0\right). \tag{35.24}$$

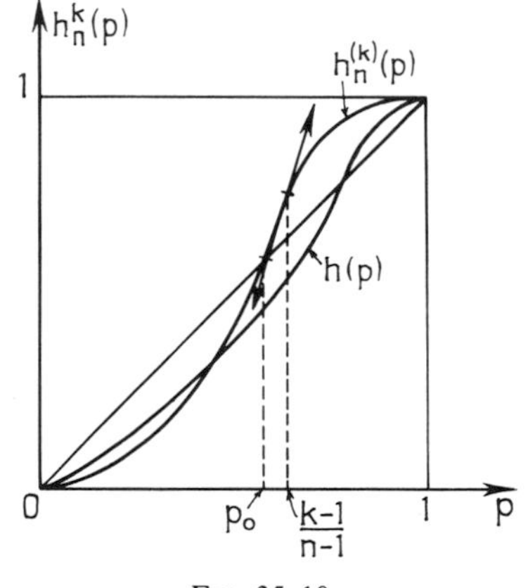

FIG. 35.10.

Figure 35.10 illustrates these properties. We have also drawn in Fig. 35.10 the reliability function of an arbitrary monotone structure, in order to illustrate the following theorem.

Theorem 35.I. *Let $h(p)$ be the reliability function of an arbitrary monotone structure of order n; if $h(p)$ is not of type k of n, the curve representative of $h(p)$ intersects at most once each curve $h_n^k(p)$, and at the point of intersection (if it exists), the slope of $h_n^k(p)$ is greater than that of $h(p)$.*

Indeed, according to (25.20), we have

$$h(p) = \sum_{i=0}^{n} A_i \, p^i (1 - p)^{n-i}, \tag{35.25}$$

where A_i is the number of links having i components. It then follows that

(35.26)

$$h_n^k(p) - h(p) = - \sum_{i=0}^{k-1} A_i p^i (1 - p)^{n-i} + \sum_{i=k}^{n} \left(\binom{n}{i} - A_i \right) p^i (1 - p)^{n-i}$$

$$= (1 - p)^n g(z) ,$$

where $g(z)$ is a polynomial of the positive variable $z = p/(1 - p)$. Since $\binom{n}{i} \geqslant A_i$, the coefficients of the polynomial $g(z)$ present at most one change of sign and $g(z)$ therefore has at most one positive root (Descartes' rule).

If $g(z)$ has a root, then $A_i > 0$ for at least one value of i less than k; but in this case if p is an infinitely small positive number, we obtain

(35.27) $$(p \to 0) \Rightarrow (h_n^k(p) < h(p)) .$$

The curve $h_n^k(p)$ therefore intersects the curve $h(p)$ from below (see Fig. 35.10).

CHAPTER VI

SYSTEMS PRESENTING TWO DUAL TYPES OF FAILURES

36 Introduction

The study of the reliability of systems of electric relays presents a particular difficulty: indeed, a relay must, according to the command sent to it, let pass or interrupt an electric current; it may thus present two types of contradictory failures:

failure of type a: circuit remains open whatever the command (no current passes);

failure of type b: circuit remains closed whatever the command (current always passes).

For example, arrange four relays according to the scheme shown in Fig. 36.1 and send the same command to all four relays. If relays 1 and 2 or 3 and 4 present failures of type b, current passes whatever the command. Likewise, if relays 1 and 3 or 1 and 4 or 2 and 3 or 2 and 4 present failures of type a, the set has failed; in fact, whatever the command, no current will be able to pass.

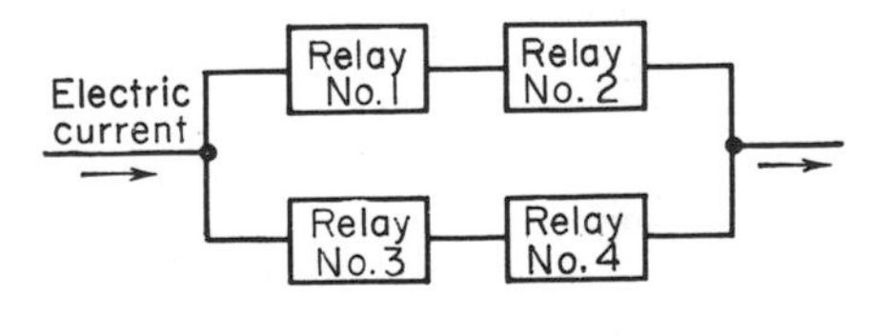

FIG. 36.1.

In this example, each component has three possible states: good state, failure a, or failure b. Similarly, the system has three possible states, defined in the same fashion. One might imagine generalizing the theory of structure functions, developed in Chapter III, to the case of systems and components presenting more than two possible states. We go on to see, however, that the systems analogous to those of the example above may, because of their peculiarities, be studied with the aid of bivalent structure functions like those discussed in the preceding chapters.

One may consider the system represented in Fig. 36.1 as serving two functions:

Function 1: when the command to close is sent, the circuit is to be closed;

Function 2: when the command to open is sent, the circuit is to be opened.

Function 1 is assured if and only if there exists in the graph formed by the electric circuit a path such that all the relays located on the path are closed; this will happen if each of the relays is either in a good state (in which case it will obey the command to close) or presents a failure of type b (since then it is always closed). Moreover, note that when function 1 is assured, the system is in a good state or presents a type b failure. One sees that this analysis takes in the three possible states of the system and of its components through only a partition into two subsets of states:

good state or failure of type b;
failure of type a.

This reduces us to taking account of only failures of type a, the failures of type b not being distinguished from "good state"; proceeding thusly, one may trace a reliability network (which will turn out to be identical to the graph formed by the electric circuit) or define a structure function.

We now pass to function 2. This function is assured if there exists a cut in the graph of the circuit such that any relay of this cut is open, that is, either in a good state or presents a type a failure. In this case, the system itself is in a good state or presents a type a failure. This second analysis requires taking into account only failures of type b, the failures of type a having been joined with "good states."

In conclusion, we verify that the state of the system as a function of the states of its components may be represented completely by *two* reliability functions or structure functions of the type we have defined in Chapter III. Moreover, these two networks or structure functions are *dual* to one another. In fact, we have seen:

that the reliability network obtained by considering only failures of type a (first analysis) coincide with the graph of the circuit;

that the reliability network obtained by considering only failures of type

b (second analysis) has for links the cuts of this graph, that is, the cuts of the first network. This second network is thus the dual of the first (Section 19, p. 79)

We shall say that the system presents two "dual" types of failures. This situation, although a very particular one, is not uncommon in the study of logical or numerical circuits, whose increasing importance in modern technology needs hardly be emphasized. This is why we shall study this in detail in the present chapter.

First we shall give another example. Many electronic components (diodes, resistors, ...) present essentially two type of failures: short circuits (zero resistance) or open circuits (infinite resistance). If one is concerned with the sudden appearance of these two types of failure between two points of a complex circuit, as was, for example, Price [45] in the particular case of components in parallel (from the point of view of electronic circuits), the method that we shall go on to describe is useful. Note, however, that the study of logic circuits raises a number of other problems that we shall not mention (see, for example, Wilcox and Mann [58]).

37 Definition. Properties

Consider a system with n components that present two types of mutually exclusive failures:

failure of type a,
failure of type b.

To each component there corresponds two Boolean state variables:

$x_i^a = 0$ if component e_i presents failure a;
$x_i^a = 1$ if component e_i is in a good state or presents failure b;
$x_i^b = 0$ if component e_i presents failure b,
$x_i^b = 1$ if component e_i is in a good state or presents failure a.

The system itself presents two types of failure, also designated a and b, coming only from, respectively, the type a and type b failures of its components. In other words, the state of the system is defined by the two following structure functions:

$\varphi(x_1^a, x_2^a, \ldots, x_n^a) = 0$ if the system presents failure a,
$= 1$ if the system does not present failure a;
$\bar{\varphi}(x_1^b, x_2^b, \ldots, x_n^b) = 0$ if the system presents failure b,
$= 1$ if the system does not present failure b.

In addition, $\bar{\varphi}(x_1^b, x_2^b, \ldots, x_n^b)$ is the dual of $\varphi(x_1^a, x_2^a, \ldots, x_n^a)$, that is,

$$\bar{\varphi}(x_1, x_2, \ldots, x_n) = 1 - \varphi(1 - x_1, 1 - x_2, \ldots, 1 - x_n) . \tag{37.1}$$

The structure function of the system, defined as taking the value 1 if the system is in a good state and the value 0 if it presents one of the two types of failure, is then

$$
\begin{aligned}
(37.2)\quad \gamma(x_1^a, x_2^a, \ldots, x_n^a; x_1^b, x_2^b, \ldots, x_n^b) \\
&= \varphi(x_1^a, x_2^a, \ldots, x_n^a) . \bar{\varphi}(x_1^b, x_2^b, \ldots, x_n^b) \\
&= \varphi(x_1^a, x_2^a, \ldots, x_n^a) . [1 - \varphi(1 - x_1^b, 1 - x_2^b, \ldots, 1 - x_n^b)] .
\end{aligned}
$$

Theorem 37.I. *The system cannot simultaneously present failures a and b, that is, one may not have* $\varphi(x^a) = \bar{\varphi}(x^b) = 0$.

Indeed, if $\bar{\varphi}(x^b) = 0$, the components presenting failure b form a cut of $\bar{\varphi}(x^b)$, but this cut of $\bar{\varphi}(x^b)$ is a link of $\varphi(x^a)$, and for all components of this link, $x_i^a = 1$, therefore $\varphi(x^a) = 1$. The system therefore has only three possible states, which we shall designate, respectively, by A (failure a), B (failure b), and C (good state).

Theorem 37.II. *If* $\varphi(x^a)$ *is monotone and nondegenerate, the system cannot function when all the components have failed.*

Indeed, if the system functions, there exist:

a link of $\varphi(x^a)$ for which all components present failure b (they have all failed, and none presents failure a);

a link of $\bar{\varphi}(x^b)$, thus a cut of $\varphi(x^a)$, for which all components present failure a.

This is impossible, however, since a path and a cut of a monotone structure have at least one component in common (Theorem 19.II on reliability networks).

Example. Consider the system of three relays arranged as shown in Fig. 37.1. For failures of type a (open circuit), the reliability network $\mathfrak{R}$ is like the electric network (Fig. 37.2). For failures of type b (closed circuit), we obtain the reliability network $\bar{\mathfrak{R}}$ by taking the dual of the network $\mathfrak{R}$. That is, the minimal links of $\bar{\mathfrak{R}}$ are the minimal cuts of $\mathfrak{R}$: $\{ e_1 \}$ and $\{ e_2, e_3 \}$.

The reliability network $\bar{\mathfrak{R}}$ is very different from the electric network (Fig. 37.3).

The structure function for $\mathfrak{R}$ is

$$\varphi(x^a) = x_1^a[1 - (1 - x_2^a)(1 - x_3^a)]$$

and that of $\bar{\mathfrak{R}}$ is

$$\bar{\varphi}(x^b) = 1 - (1 - x_1^b)(1 - x_2^b x_3^b)$$

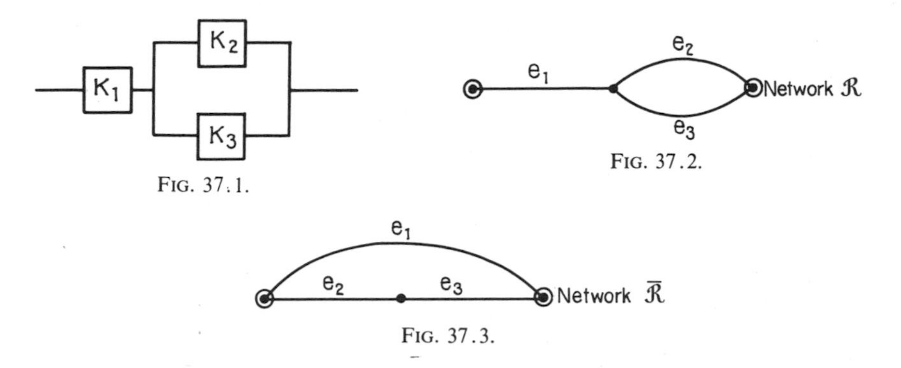

FIG. 37.1.

FIG. 37.2.

FIG. 37.3.

and that of the system of relays is

$$\gamma(x^a, x^b) = \varphi(x^a)\, \overline{\varphi}(x^b) .$$

We may with this example verify Theorem 37.II. If, for example, relay K_2 has failure a and relays K_1 and K_3 have failure b, we obtain

$$\begin{aligned} x_1^a &= 1 , & x_1^b &= 0 , \\ x_2^a &= 0 , & x_2^b &= 1 , \\ x_3^a &= 1 , & x_3^b &= 0 , \end{aligned}$$

and it follows that

$$\gamma(x^a, x^b) = [1 - (1 - 0)(1 - 1)]\, [1 - (1 - 0)(1 - 1 \times 0)] = 0 .$$

38 Reliability Function of a System Presenting Two Types of Failures

Definition. Denote by

p_i^a the probability that component e_i has failure a,
p_i^b the probability that component e_i has failure b.

From this

$$1 - h(1 - p_1^a, 1 - p_2^a, \ldots, 1 - p_n^a) ,$$

the probability that the system is in state A (presents failure a),

$$1 - \overline{h}(1 - p_1^b, 1 - p_2^b, \ldots, 1 - p_n^b) ,$$

the probability that the system is in state B (presents failure b).

We obviously have

$$1 - h(1 - p^a) = 1 - \varphi_s(1 - p^a) \tag{38.1}$$

where $\varphi_s(x)$ is the simple form (see 31.5)) of the structure function φ, and

(38.2)

$$1 - \bar{h}(1 - p^b) = 1 - \bar{\varphi}_s(1 - p^b) = 1 - [1 - \varphi_s(p^b)] = \varphi_s(p^b) = h(p^b) .$$

Denoting by C the state of the system presenting neither a nor b failures, we must have

$$\text{prob}\{ A \} + \text{prob}\{ B \} + \text{prob}\{ C \} = 1 \tag{38.3}$$

since events A, B, and C are incompatible (see Theorem 37.I).

Writing $f(p_1^a, p_2^a, \ldots, p_n^a; p_1^b, p_2^b, \ldots, p_n^b)$ for the reliability function of the system, we have

(38.4)

$$\begin{aligned}\text{prob}\{ C \} &= f(p_1^a, p_2^a, \ldots, p_n^a ; p_1^b, p_2^b, \ldots, p_n^b)\\ &= 1 - [1 - h(1 - p_1^a, 1 - p_2^a, \ldots, 1 - p_n^a)] - h(p_1^b, p_2^b, \ldots, p_n^b)\end{aligned}$$

or

(38.5)

$$\begin{aligned}&f(p_1^a, p_2^a, \ldots, p_n^a ; p_1^b, p_2^b, \ldots, p_n^b)\\ &\qquad = h(1 - p_1^a, 1 - p_2^a, \ldots, 1 - p_n^a) - h(p_1^b, p_2^b, \ldots, p_n^b)\\ &\qquad = \varphi_s(1 - p_1^a, 1 - p_2^a, \ldots, 1 - p_n^a) - \varphi_s(p_1^b, p_2^b, \ldots, p_n^b) .\end{aligned}$$

Example 1. Consider a system of relays in series–parallel (Fig. 38.1), each relay receiving the same command (see Section 36).

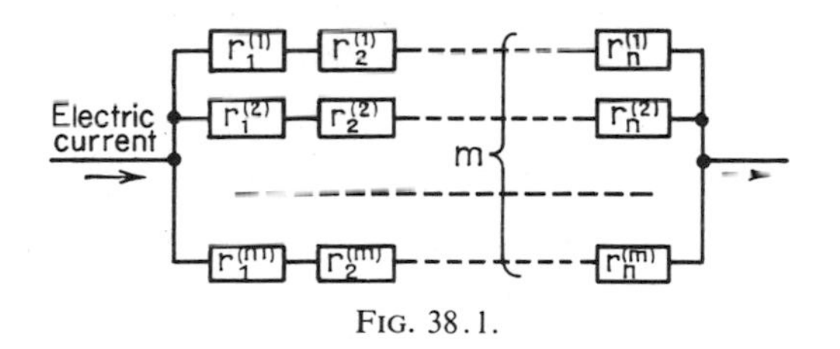

FIG. 38.1.

Let a be the "circuit always open" failure; any series of n relays $\{ r_1^{(j)}, r_2^{(j)}, \ldots, r_n^{(j)} \}$ is a link of $\varphi(x_1^a, x_2^a, \ldots, x_n^a)$ and we obtain

$$\varphi_s(x_1^a, x_2^a, \ldots, x_n^a) = 1 - \prod (1 - x_1^a x_2^a \ldots x_n^a) \tag{38.6}$$

where the product is taken over m minimal links.

Suppose that the $m \times n$ relays all have the same reliability. Let α be the probability of a type a failure (circuit always open) and β be the probability of a failure of type b (circuit always closed); one has

$$h(\alpha) = 1 - (1 - \alpha^n)^m \tag{38.7}$$

and

$$f(\alpha, \beta) = (1 - \beta^n)^m - [1 - (1 - \alpha)^n]^m . \tag{38.8}$$

Example 2. Consider the system of five relays shown in Fig. 38.2. As in the preceding example, let α and β be the probabilities of failures of type a (circuit always open) and type b (circuit always closed).

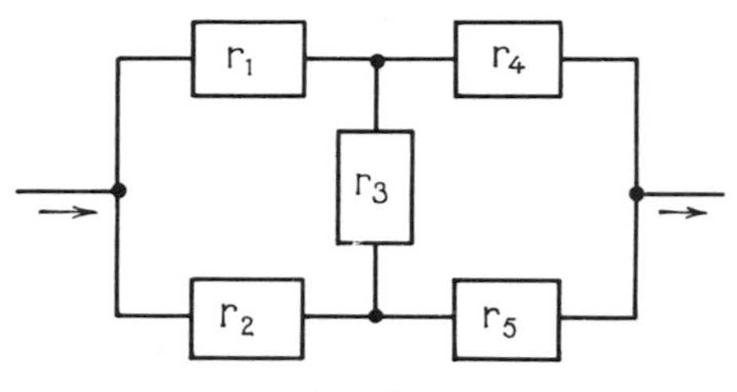

FIG. 38.2.

The minimal links of the structure $\varphi(x_1^a, x_2^a, x_3^a, x_4^a, x_5^a)$ are

$$\{ r_1, r_4 \}, \quad \{ r_2, r_5 \}, \quad \{ r_1, r_3, r_5 \}, \text{ and } \{ r_2, r_3, r_4 \}, \tag{38.9}$$

and from this

(38.10)

$$\varphi(x_1^a, x_2^a, x_3^a, x_4^a, x_5^a) = 1 - (1 - x_1^a x_4^a)(1 - x_2^a x_5^a) \cdot (1 - x_1^a x_3^a x_5^a)(1 - x_2^a x_3^a x_4^a),$$

from which

(38.11)

$$\begin{aligned} \varphi_s(x_1^a, x_2^a, x_3^a, x_4^a, x_5^a) = {} & x_1^a x_4^a + x_2^a x_5^a + x_1^a x_3^a x_5^a + x_2^a x_3^a x_4^a \\ & - x_1^a x_2^a x_4^a x_5^a - x_1^a x_2^a x_3^a x_4^a - x_1^a x_2^a x_3^a x_5^a \\ & - x_1^a x_3^a x_4^a x_5^a - x_2^a x_3^a x_4^a x_5^a + 2\, x_1^a x_2^a x_3^a x_4^a x_5^a \end{aligned}$$

and

$$h(\alpha) = 2\,\alpha^2 + 2\,\alpha^3 - 5\,\alpha^4 + 2\,\alpha^5 , \tag{38.12}$$

and thus finally

(38.13)

$$f(\alpha, \beta) = (1 - \alpha)^2 [4 - 2\,\alpha - 5(1 - \alpha)^2 + 2(1 - \alpha)^3] - \beta^2[2 + 2\,\beta - 5\,\beta^2 + 2\,\beta^3] .$$

Moore–Shannon Upper Bound. As in the case of a system with only one type of failure (Section 25), one may indicate limits for the reliability function, as a function of the length μ and the width λ of the system, whenever the components are all identical. Recall (Section 17, p. 66) that the "length" of a system is the smallest number of components forming a link, and the "width" the smallest number forming a cut. Here we shall apply these definitions considering only failures of type a (the second reliability network, obtained by considering only failures of type b, is dual to the first, and has therefore length μ and width λ).

If all the components have the same probabilities of failure α (type a) and β (type b), one has, from Theorem 25.V (p. 123),

$$\beta^{\lambda} \leqslant h(\beta) \leqslant 1 - (1 - \beta)^{\mu} \tag{38.14}$$

and likewise

$$(1 - \alpha)^{\lambda} \leqslant h(1 - \alpha) \leqslant 1 - \alpha^{\mu} . \tag{38.15}$$

The reliability function of the system, given by

$$f(\alpha, \beta) = h(1 - \alpha) - h(\beta) , \tag{38.16}$$

therefore satisfies the inequalities

$$(1 - \alpha)^{\lambda} + (1 - \beta)^{\mu} - 1 \leqslant f(\alpha, \beta) \leqslant 1 - \alpha^{\mu} - \beta^{\lambda} . \tag{38.17}$$

Theorem 38.I. *In order that the probability of failure (type a) of the system be less than δ_1 and the probability of failure (type b) be less than δ_2, it is necessary that the number of components n satisfy the inequality*

$$n \geqslant \frac{\log \delta_1}{\log \alpha} \cdot \frac{\log \delta_2}{\log \beta} \tag{38.18}$$

where α is the probability of type a failure of each of the components, and β is the probability of type b failure of each of the components.

Indeed, the probabilities of type a and of type b failures must satisfy the inequalities

$$\alpha^{\mu} \leqslant 1 - h(1 - \alpha) \leqslant \delta_1 , \tag{38.19}$$

$$\beta^{\lambda} \leqslant h(\beta) \leqslant \delta_2 , \tag{38.20}$$

and this entails

$$\mu \log \alpha \leqslant \log \delta_1 , \tag{38.21}$$

$$\lambda \log \beta \leqslant \log \delta_2 . \tag{38.22}$$

Since $\log \alpha$ and $\log \beta$ are negative ($\alpha < 1$ and $\beta < 1$), we obtain

$$\lambda \geqslant \frac{\log \delta_2}{\log \beta} ,$$

$$\mu \geqslant \frac{\log \delta_1}{\log \alpha} .$$

By applying Theorem 19.VI, p. 77, one may write

$$n \geqslant \lambda.\mu \geqslant \frac{\log \delta_1}{\log \alpha} \cdot \frac{\log \delta_2}{\log \beta} .$$

39 Redundance in Systems with Components of the Same Reliabilities

In the case studied here of two types of failure, the optimal procedure for redundance depends on the relative frequencies of each type of failure. If, for example, it is a matter of juxtaposing a redundant component to a given component, and if almost all failures are of type a (circuit always open), the most effective redundance consists in arranging the two components in parallel; if, on the other hand, the most frequent failure is of type b (circuit always closed), it would be better to arrange the two components in series.

We shall give in succession a few results concerning the simplest monotone structures (orders 1, 2, and 3), and series–parallel networks. In the following section we shall examine a particular kind of redundance, composition of a structure with itself.

In the present section we shall suppose that all the components have the same reliabilities, α being the probability of type a failure and β the probability of type b failure.

Simple Monotone Structures. Optimal Arrangement of One or Two Redundant Components. If we wish to arrange a redundant component, we may place it in parallel or in series with the initial component of the structure; we obtain, according to (26.58):

For the parallel arrangement (Fig. 39.1):

$$h_1^{(2)}(\alpha) = 2\,\alpha - \alpha^2 . \tag{39.1}$$

For the series arrangement (Fig. 39.2):

$$h_2^{(2)}(\alpha) = \alpha^2 . \tag{39.2}$$

FIG. 39.1.

FIG. 39.2.

In order to study the effectiveness of these two arrangements, we thus must compare

$$f_1^{(2)}(\alpha, \beta) = 2(1 - \alpha) - (1 - \alpha)^2 - 2\beta + \beta^2 \tag{39.3}$$

and

$$f_2^{(2)}(\alpha, \beta) = (1 - \alpha)^2 - \beta^2 . \tag{39.4}$$

For this, we introduce a new variable

$$\gamma = 1 - \alpha - \beta , \tag{39.5}$$

which represents the reliability of a component (neither type *a* nor type *b* failure). Note that $\gamma \geqslant 0$, from expression (38.3).

We have

$$f_1^{(2)}(\alpha, \beta) = 2\gamma - \gamma(\gamma + 2\beta) , \tag{39.6}$$

$$f_2^{(2)}(\alpha, \beta) = \gamma(\gamma + 2\beta) . \tag{39.7}$$

The relative increment in reliability is therefore

$$\frac{f_1^{(2)}(\alpha, \beta)}{\gamma} = 2 - \gamma - 2\beta , \tag{39.8}$$

$$\frac{f_2^{(2)}(\alpha, \beta)}{\gamma} = \gamma + 2\beta . \tag{39.9}$$

Then we obtain the graph of Fig. 39.3.

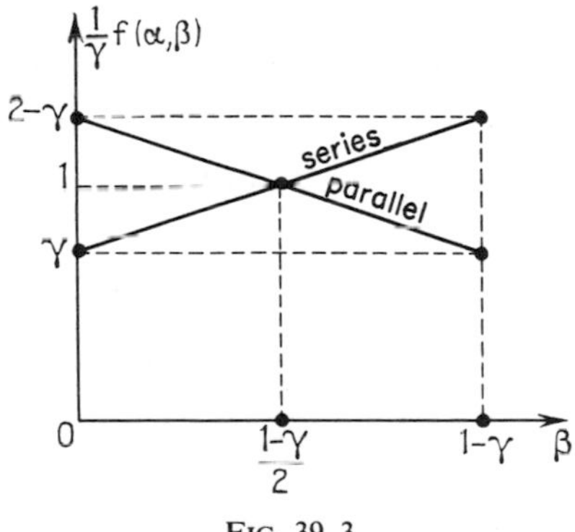

FIG. 39.3.

Note that since $h_2^{(2)}$ is the dual structure of $h_1^{(2)}$, $f_2^{(2)}$ follows from $f_1^{(2)}$ by permuting α and β and the two lines are symmetrical with respect to the line

$$\beta = \frac{1 - \gamma}{2} = \alpha .$$

When $\beta < (1 - \gamma)/2$, and therefore $\beta < \alpha$, the parallel arrangement is more effective; conversely, it is the series arrangement that is more effective if $\beta > (1 - \gamma)/2$; thus $\beta > \alpha$. If $\beta = (1 - \gamma)/2$, that is, $\beta = \alpha$, redundance is useless.

We now consider the case of two redundant components. We shall use the results given in Section 26 on monotone structures of order 3 ((26.60)–(26.64) and Fig. 26.13). We obtain the following results:

(39.10)

$$\begin{aligned} f_1^{(3)}(\alpha, \beta) &= 3(1 - \alpha) - 3(1 - \alpha)^2 \\ &\quad + (1 - \alpha)^3 - 3\,\beta + 3\,\beta^2 - \beta^3 \\ &= (1 - \beta)^3 - \alpha^3 ; \end{aligned}$$

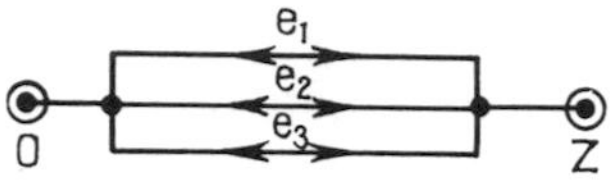

e1 e2 e3 O Z

(39.11)

$$\begin{aligned} f_2^{(3)}(\alpha, \beta) &= 1 - \alpha + (1 - \alpha)^2 - (1 - \alpha)^3 \\ &\quad - \beta - \beta^2 + \beta^3 \\ &= 1 - 2\,\alpha^2 + \alpha^3 - \beta - \beta^2 + \beta^3 ; \end{aligned}$$

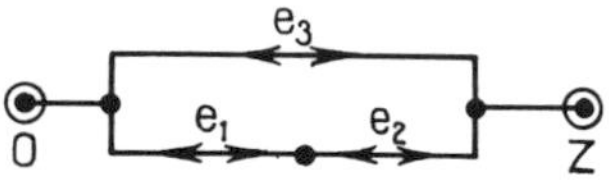

e3 e1 e2 O Z

(39.12)

$$\begin{aligned} f_4^{(3)}(\alpha, \beta) &= 3(1 - \alpha)^2 - 2(1 - \alpha)^3 \\ &\quad - 3\,\beta^2 + 2\,\beta^3 \\ &= 1 - 3\,\alpha^2 + 2\,\alpha^3 - 3\,\beta^2 + 2\,\beta^3 ; \end{aligned}$$

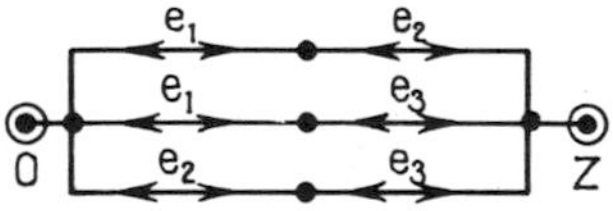

e1 e2 e1 e3 e2 e3 O Z

(39.13)

$$\begin{aligned} f_5^{(3)}(\alpha, \beta) &= 2(1 - \alpha)^2 - (1 - \alpha)^3 \\ &\quad - 2\,\beta^2 + \beta^3 \\ &= 1 - \alpha - \alpha^2 + \alpha^3 - 2\,\beta^2 + \beta^3 ; \end{aligned}$$

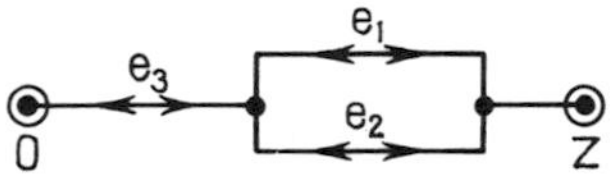

e1 e3 e2 O Z

(39.14)

$$f_7^{(3)}(\alpha, \beta) = (1 - \alpha)^3 - \beta^3 .$$

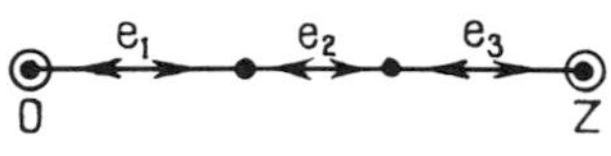

e1 e2 e3 O Z

The graph given in Fig. 39.4 is obtained. We have set

$$\gamma = 1 - \alpha - \beta ,$$

and we have used as ordinate $(1/\gamma) f(\alpha, \beta)$, which is the relative increment in reliability. Note that the graph is symmetric with respect to the line

$$\beta = \frac{1 - \gamma}{2} .$$

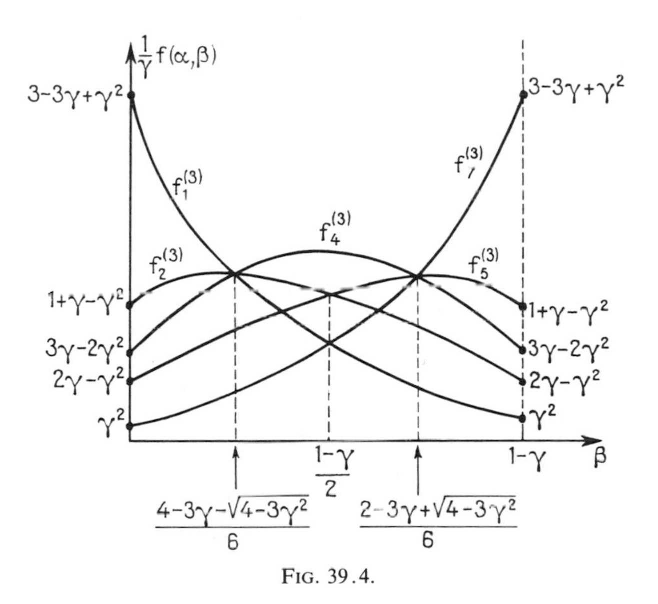

FIG. 39.4.

Note also that $f_1^{(3)}$, $f_2^{(3)}$, and $f_4^{(3)}$ have a common point of intersection, and similarly for f_4^3, f_5^3, and f_7^3.

From Fig. 39.4 we draw the following conclusions: If

$$\beta \leqslant \frac{4 - 3\gamma - \sqrt{4 - 3\gamma^2}}{6}, \tag{39.15}$$

the arrangement $f_1^{(3)}$ is most effective. If

$$\frac{4 - 3\gamma - \sqrt{4 - 3\gamma^2}}{6} \leqslant \beta \leqslant \frac{2 - 3\gamma + \sqrt{4 - 3\gamma^2}}{6}, \tag{39.16}$$

one would prefer $f_4^{(3)}$. If

$$\beta \geqslant \frac{2 - 3\gamma + \sqrt{4 - 3\gamma^2}}{6}, \tag{39.17}$$

one would prefer $f_7^{(3)}$.

At the point

$$\beta = \frac{4 - 3\gamma - \sqrt{4 - 3\gamma^2}}{6},$$

$f_1^{(3)}$, $f_2^{(3)}$, $f_4^{(3)}$ are equal, but these structures are preferable to $f_5^{(3)}$ and to $f_7^{(3)}$.

At the point

$$\beta = \frac{2 - 3\gamma + \sqrt{4 - 3\gamma^2}}{6},$$

$f_4^{(3)}, f_5^{(3)}, f_7^{(3)}$ are equal, but these structures are preferable to $f_1^{(3)}$ and to $f_2^{(3)}$.

Series–Parallel Structures. Consider the structure represented in Fig. 39.5 and make the hypothesis that all the components possess the same probabilities α and β.

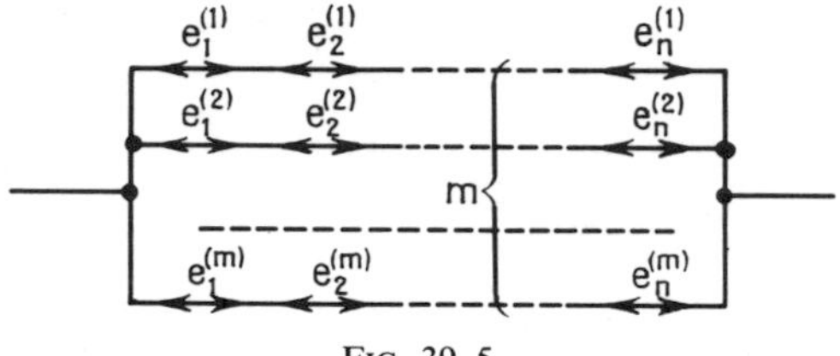

FIG. 39.5.

The reliability of this structure is (cf. (38.8))

$$f(\alpha, \beta) = (1 - \beta^n)^m - [1 - (1 - \alpha)^n]^m . \tag{39.18}$$

With n fixed, we denote by m^* the value of m that maximizes the reliability of the series–parallel structure.

We may easily show (see Barlow and Proschan [5]) that m^* has the following properties:

Property 1.

$$m^* = [m_0] + 1 \tag{39.19}$$

where $[X]$ signifies the "integer part of" X, with

$$m_0 = n \frac{\log(1 - \alpha) - \log\beta}{\log(1 - \beta^n) - \log[1 - (1 - \alpha)^n]} . \tag{39.20}$$

Property 2.

(39.21) m_0/n is an increasing function of n.

Property 3. If $n \to \infty$,

$$\frac{m_0}{n} \simeq \left(\frac{1}{1 - \alpha}\right)^n \ln\frac{1 - \alpha}{\beta} . \tag{39.22}$$

Property 4.

(39.23) $$(\beta^n + (1-\alpha)^n > 1) \Rightarrow (m^* = 1),$$

(39.24) $$(\beta^n + (1-\alpha)^n = 1) \Rightarrow (m^* = 1 \text{ or } 2),$$

(39.25) $$(\beta^n + (1-\alpha)^n < 1) \Rightarrow (m^* > 1).$$

40 Iterative Structures

We first define the notion of iterative structure in the case of systems with only one type of failure, using the composition of structures defined in Section 26.

Let $\varphi(x_1, x_2, \dots, x_n)$ be a monotone structure function. Let us suppose that these Boolean quantities are themselves structure functions:

(40.1)
$$\begin{aligned} x_1 &= \varphi(x_{11}, x_{12}, \dots, x_{1n}), \\ x_2 &= \varphi(x_{21}, x_{22}, \dots, x_{2n}), \\ &\dots, \\ x_n &= \varphi(x_{n1}, x_{n2}, \dots, x_{nn}). \end{aligned}$$

Put

(40.2)
$$\begin{aligned} (u_1) &= (x_{11}, x_{12}, \dots, x_{1n}), \\ (u_2) &= (x_{21}, x_{22}, \dots, x_{2n}), \\ &\dots, \\ (u_n) &= (x_{n1}, x_{n2}, \dots, x_{nn}). \end{aligned}$$

Make the hypothesis that these n n-tuples are independent, that is, that the corresponding components form modules (see Section 26); we then put

(40.3)
$$\begin{aligned} \varphi_2((u_1), (u_2), \dots, (u_n)) &= \varphi[\varphi(u_1), \varphi(u_2), \dots, \varphi(u_n)] \\ &= \varphi[\varphi(x_{11}, x_{12}, \dots, x_{1n}), \varphi(x_{21}, x_{22}, \dots, x_{2n}), \\ &\qquad \dots, \varphi(x_{n1}, x_{n2}, \dots, x_{nn})]. \end{aligned}$$

Thus we have realized a composition of the structure function φ in itself.

Example. Consider the network of Fig. 40.1; the monotone structure function that corresponds to it is (cf. Fig. 26.13, p. 134)

(40.4) $$\varphi(x_1, x_2, x_3) = x_1 x_3 + x_2 x_3 - x_1 x_2 x_3.$$

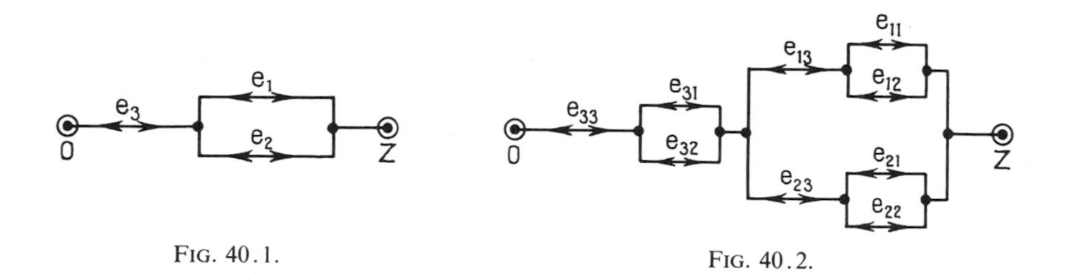

FIG. 40.1.

FIG. 40.2.

Now, suppose that each component e_1, e_2, e_3 is itself a module having the same structure as that of the network in Fig. 40.1. By considering the $3 \times 3 = 9$ components that occur in the new network (Fig. 40.2) we may write

(40.5)

$$\begin{aligned}
&\varphi_2(x_{11}, x_{12}, x_{13}, x_{21}, x_{22}, x_{23}, x_{31}, x_{32}, x_{33}) \\
&= (x_{11}\, x_{13} + x_{12}\, x_{13} - x_{11}\, x_{12}\, x_{13})\,(x_{31}\, x_{33} + x_{32}\, x_{33} - x_{31}\, x_{32}\, x_{33}) \\
&\quad + (x_{21}\, x_{23} + x_{22}\, x_{23} - x_{21}\, x_{22}\, x_{23})\,(x_{31}\, x_{33} + x_{32}\, x_{33} - x_{31}\, x_{32}\, x_{33}) \\
&\quad - (x_{11}\, x_{13} + x_{12}\, x_{13} - x_{11}\, x_{12}\, x_{13})\,(x_{21}\, x_{23} + x_{22}\, x_{23} - x_{21}\, x_{22}\, x_{23}) \\
&\quad . (x_{31}\, x_{33} + x_{32}\, x_{33} - x_{31}\, x_{32}\, x_{33}) \, .
\end{aligned}$$

We leave this exercise to return to the general exposition. Suppose that all the components have the same reliability p; we may then write, by considering the corresponding reliability functions,

(40.6) $$h_2(p) = h(h(p)) \, .$$

Indeed, if the structure function φ is put in simple form, expression (40.3) similarly gives φ_2 a simple form since it corresponds to a composition of modules (see the example above); the reliability functions are then connected by the same relations. Similarly,

(40.7) $$h_3(p) = h[h_2(p)] = h[h\{\, h(p)\,\}] \, ,$$

$$\ldots,$$

(40.8) $$h_n(p) = h[h_{n-1}(p)] \, .$$

Example. Consider again the example of Figs. 40.1 and 40.2. We have

(40.9) $$h(p) = 2\, p^2 - p^3 \, ,$$

from which

$$(40.10)\quad \begin{aligned} h_2(p) &= 2[h(p)]^2 - [h(p)]^3 \\ &= 2(2\,p^2 - p^3)^2 - (2\,p^2 - p^3)^3 \\ &= (2 - 2\,p^2 + p^3)\,(2\,p^2 - p^3)^2 , \end{aligned}$$

$$(40.11)\quad \begin{aligned} h_3(p) &= h_2[h(p)] \\ &= (2 - 2[h(p)]^2 + [h(p)]^3)\,(2[h(p)]^2 - [h(p)]^3)^2 \\ &= [2 - 2(2\,p^2 - p^3)^2 + (2\,p^2 - p^3)^3] \\ &\qquad \cdot [2(2\,p^2 - p^3)^2 - (2\,p^2 - p^3)^3]^2 . \end{aligned}$$

We return to the general case and graphically represent the realized iterations using the curves studied in Section 27 (Fig. 27.2). We see that (Fig. 40.3):

(1) if the reliability function is of type I, the iteration diminishes the reliability of the set;

(2) if the reliability function is of type II, the iteration increases the reliability whenever $p > p_0$ and diminishes it when $p < p_0$, where p_0 is the nonzero solution of $h(p) = p$;

(3) if the reliability function is of type III, iteration increases the reliability.

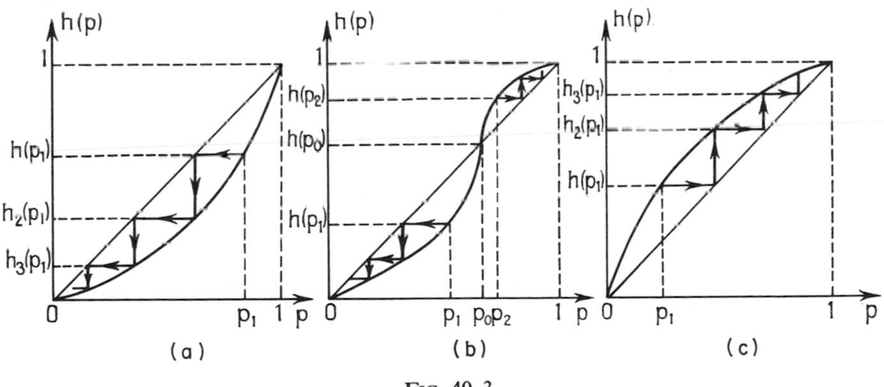

FIG. 40.3.

Now we go on to the case where the system possesses two types of failures. Then, as we have seen, the reliability function of the system is

$$(40.12)\qquad f(\alpha, \beta) = h(1 - \alpha) - h(\beta) .$$

If we desire to improve the reliability, we must increase $h(1 - \alpha)$ and reduce $h(\beta)$. To reach this goal through an iterative composition, we must have:

(1) on one hand, that the reliability function $h(p)$ be of type II,
(2) on the other hand,

$$\beta < p_0 < 1 - \alpha \,, \tag{40.13}$$

p_0 being the solution of the equation

$$h(p) = p \,. \tag{40.14}$$

A simple solution to this problem of improving the reliability consists in choosing a k of n structure (see Section 35) for which the reliability function cuts the first bisector at a point p_0 satisfying (40.13).

Birnbaum and co-workers [8] have calculated a table of values of p_0 for $n \leqslant 25$. In Table 40.I we give an extract from this table.

p_0	0.50	0.40	0.35	0.26	0.23	0.13	0.08	0.025	0.01	0.05
n	3	8	6	7	4	5	6	10	15	22
k	2	4	3	3	2	2	2	2	2	2

TABLE 40.I

Moore and Shannon give a general method, somewhat more complex, for increasing the reliability in the case considered here.

APPENDIX

PÓLYA FUNCTIONS OF ORDER 2. TOTALLY POSITIVE FUNCTIONS OF ORDER 2

A.1 Pólya Functions of Order 2

Definition *A function $f(x)$ defined in $\mathbf{R}$, taking its values in $\mathbf{R}^+ = [0, \infty]$, and such that*

(A1.1) $\forall\, x_1, x_2, y_1, y_2 \in \mathbf{R}$, *with* $x_1 \leqslant x_2, y_1 \leqslant y_2$:

$$\begin{vmatrix} f(x_1 - y_1) & f(x_1 - y_2) \\ f(x_2 - y_1) & f(x_2 - y_2) \end{vmatrix} \geqslant 0$$

is called a Pólya function of order 2 (in the following, the phrase of order 2 will be understood).

Condition (A1.1) may also be written, by expanding the determinant, as

(A1.2) $$f(x_1 - y_1)\, f(x_2 - y_2) \geqslant f(x_1 - y_2)\, f(x_2 - y_1)\,.$$

The four arguments that occur in this inequality have, from the conditions $x_1 \leqslant x_2$ and $y_1 \leqslant y_2$, relative positions on the real axis as shown schematically by the lattice of Fig. A1.1, where the arrows point from each quantity to one that is less than or equal to the first.

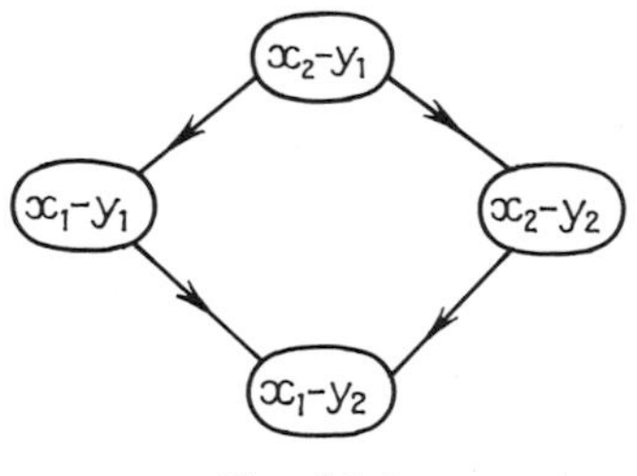

FIG. A1.1.

Some Properties

(1) *If $f(x)$ is a Pólya function, then the function*

$$g(x) = f(-x) \tag{A1.3}$$

is also a Pólya function.

This property results from the interchangeability of the pairs (x_1, x_2) and (y_1, y_2) in definition (A1.2).

(2) *If $f(x)$ is a Pólya function, the set*

$$\mathbf{I} = \{ x / f(x) > 0 \} \tag{A1.4}$$

is an interval.

We set aside the trivial case of an identically zero function, where $\mathbf{I} = \varnothing$.

Conversely, if $f(x)$ is nonzero, $\mathbf{I} = \mathbf{R}$. Suppose now that $f(x)$ is not identically zero, but is made zero by at least one value x_0 of x. We apply (A1.1) in the particular case where

$$\begin{aligned} x_1 &= x_0 , \qquad & x_2 &> x_0 , \\ y_1 &= 0 , \qquad & y_2 &> 0 . \end{aligned} \tag{A1.5}$$

It follows that

$$f(x_0) . f(x_2 - y_2) \geqslant f(x_0 - y_2) . f(x_2),$$

that is, since $f(x_0) = 0$ and the second term is nonnegative,

$$f(x_0 - y_2) . f(x_2) = 0 . \tag{A1.6}$$

If there exists a $y_2 > 0$ such that $f(x_0 - y_2) > 0$, one must have $f(x_2) = 0$ for all $x_2 > x_0$; conversely, if there exists an $x_2 > x_0$ such that $f(x_2) > 0$, one must have $f(x_0 - y_2) = 0$ for all $y_2 > 0$. We thus see that a value x_0 of x for which $f(x)$ is zero cannot be bracketed by values of x for which $f(x) > 0$, which shows that $\mathbf{I}$ is an interval. This interval may be unbounded on the left or on the right; it may be open or closed.

(3) *A nonnegative function $f(x)$ is a Pólya function if and only if it satisfies the following two conditions:*

(a) *the set* **I** *defined by* (A1.4) *is an interval;*

(b) $f(x)$ *satisfies* (A1.2) *for* x_1, x_2, y_1, y_2 *such that* $x_1 \leqslant x_2, y_1 \leqslant y_2$, *and such that the four arguments appearing in* (A1.2) *belong to the closure of* **I**.

In other words, if $f(x) = 0$ outside of the closed interval $[a, b]$, and $f(x) > 0$ in the open interval $]a, b[$, where a may be replaced by $-\infty$ and b by $+\infty$, it suffices to satisfy (A1.2) whenever the four arguments are all in $[a, b]$, that is, for (see Fig. A1.1)

(A1.7) $$x_1 - y_2 \geqslant a \quad \text{and} \quad x_2 - y_1 \leqslant b\,.$$

This property follows immediately from the fact that if $x_1 - y_2 < a$ and/or $x_2 - y_1 > b$, the second member of (A1.2) is zero; the first term being nonnegative, (A1.2) is certainly satisfied.

(4) *A Pólya function $f(x)$ that is piecewise continuous is unimodal; that is, either it is nondecreasing, or it is nonincreasing, or there exists an x_0 such that $f(x)$ is nondecreasing for $x \leqslant x_0$ and nonincreasing for $x \geqslant x_0$. Moreover, it may be discontinuous only at the endpoints of the interval* **I**.

Indeed, let

(A1.8) $$g(x) = \ln f(x)\,.$$

The function $g(x)$ is defined in **I**. Set

(A1.9) $$x_1 - y_1 = x_2 - y_2 = \frac{(x_2 - y_1) + (x_1 - y_2)}{2}, \qquad (x_2 - y_1) \in \mathbf{I}, \quad (x_1 - y_2) \in \mathbf{I}.$$

(A1.2) may then be written as

(A1.10) $$f^2(x_1 - y_1) \geqslant f(x_1 - y_2) . f(x_2 - y_1)$$

or

(A1.11) $$g(x_1 - y_1) \geqslant \frac{g(x_1 - y_2) + g(x_2 - y_1)}{2}.$$

The value of g at the middle of an interval is therefore at least equal to the average of the values that it takes at the extremities of this interval. If $x_1 - y_2$ and $x_2 - y_1$ are rational, one may, by continuing to divide the interval into 2, 4, 8, ... equal parts, show that $g(x)$ is concave on the set of rationals. The fact that $f(x)$, and as a consequence $g(x)$, are continuous by intervals permits the extension of this property to the entire interval, but if $g(x)$ is concave, it is either nondecreasing, or nonincreasing, or there exists x_0 such that $g(x)$

is nondecreasing for $x \leqslant x_0$ and nonincreasing for $x \geqslant x_0$. Since the logarithm function is increasing, $f(x)$ has these same properties. On the other hand, the function $g(x)$, concave in **I**, is necessarily continuous there; it is therefore the same for $f(x)$, which may be discontinuous only at the endpoints of **I**.

(5) *A function* $f(x)$ *such that*

$$f(x) = C \qquad \forall x \leqslant x_0 \qquad (\text{resp. } x \geqslant x_0), \tag{A1.12}$$

where $C > 0$, *is a Pólya function if and only if:*

(a) *it is nonincreasing* (*respectively, nondecreasing*);
(b) *it satisfies* (A1.1) *for*

$$x_1 - y_2 \geqslant x_0 \qquad (\text{resp. } x_2 - y_1 \leqslant x_0). \tag{A1.13}$$

We shall restrict ourselves to the case of piecewise continuous functions. The function $g(x)$ defined by (A1.8) being constant for $x \leqslant x_0$ (respectively, $x \geqslant x_0$), the condition that it be concave implies that it is nonincreasing (respectively, nondecreasing), and it is the same for $f(x)$. Conversely, if $g(x)$ is nonincreasing (respectively, nondecreasing), and if it is concave for $x \geqslant x_0$ (respectively, $x \leqslant x_0$), it is concave in **R**, and $f(x)$ is a Pólya function.

Examples of Pólya Functions

(1) The function

$$f(x) = e^{ax}, \qquad x \in \mathbf{R}, \qquad a \in \mathbf{R}, \tag{A1.14}$$

is a Pólya function. Indeed, one has $f(x) > 0$ and

$$e^{a(x_1 - y_1)} . e^{a(x_2 - y_2)} = e^{a(x_1 - y_2)} . e^{a(x_2 - y_1)}. \tag{A1.15}$$

(2) The complementary distribution function

$$\begin{aligned} v(t) &= 1, \qquad t \leqslant 0, \\ &= e^{-at}, \qquad t \geqslant 0, \end{aligned} \tag{A1.16}$$

where $a > 0$ is a Pólya function; this follows from property (5) above, and from the fact that (A1.14) is a Pólya function (it suffices to exchange a for $-a$).

(3) The function

$$\begin{aligned} f(x) &= x\, e^{-x^2/2}, \qquad x \geqslant 0, \\ &= 0, \qquad x \leqslant 0, \end{aligned} \tag{A1.17}$$

is a Pólya function. Indeed, according to property (3), it suffices to check (A1.2) for nonnegative values of the arguments, but the inequality

(A1.18)

$$(x_1 - y_1)(x_2 - y_2)\exp\left(-\frac{(x_1 - y_1)^2 + (x_2 - y_2)^2}{2}\right)$$
$$\geqslant (x_1 - y_2)(x_2 - y_1)\exp\left(-\frac{(x_1 - y_2)^2 + (x_2 - y_1)^2}{2}\right)$$

follows from the following two inequalities, which we may easily verify:

(A1.19)

$$(x_1 - y_1)(x_2 - y_2) \geqslant (x_1 - y_2)(x_2 - y_1) \quad \text{if} \quad x_2 \geqslant x_1, \quad y_2 \geqslant y_1,$$

(A1.20)

$$(x_1 - y_1)^2 + (x_2 - y_2)^2 \leqslant (x_1 - y_2)^2 + (x_2 - y_1)^2 \quad \text{if} \quad x_2 \geqslant x_1, \quad y_2 \geqslant y_1.$$

(4) The function

(A1.21)
$$\begin{aligned} f(x) &= x + 1, & x &\geqslant -1, \\ &= 0, & x &\leqslant -1, \end{aligned}$$

is a Pólya function since the inequality

(A1.22) $$(x_1 - y_1 + 1)(x_2 - y_2 + 1) \geqslant (x_1 - y_2 + 1)(x_2 - y_1 + 1)$$

may be reduced to (A1.19).

(5) One may also verify that the complementary distribution function

(A1.23)
$$\begin{aligned} v(t) &= 1, & t &\leqslant 0, \\ &= e^{-(at)^\beta}, & t &\geqslant 0, \quad \beta \geqslant 1, \end{aligned}$$

is a Pólya function. This function generalizes (A1.19) (see the Weibull law (6.18)).

Theorem A.I. *A nonnegative function $f(x)$ is a Pólya function if and only if:*

(a) $f(x) \geqslant 0$, *and the set* $\mathbf{I} = \{ x \mid f(x) > 0 \}$ *is an interval;*
(b) $\forall h \in \mathbf{R}_0^+$, *the function of* x

(A1.24)
$$\frac{f(x - h)}{f(x)}$$

is nondecreasing in the set $\mathbf{I}$ *where it is defined.*

We have seen that any Pólya function satisfies condition (a) (definition of a Pólya function and property (2)). We show that if z_1 and z_2 are two numbers such that $z_1 \leqslant z_2$, $f(z_1) > 0$ and $f(z_2) > 0$, and h is a positive number, then we have

$$\frac{f(z_1 - h)}{f(z_1)} \leqslant \frac{f(z_2 - h)}{f(z_2)}. \tag{A1.25}$$

Choose x_2 arbitrarily and put

$$\begin{aligned} x_1 &= x_2 - h, \\ y_1 &= x_2 - z_2, \\ y_2 &= x_2 - z_1. \end{aligned} \tag{A1.26}$$

One has $x_1 \leqslant x_2$ and $y_1 \leqslant y_2$, from which

$$f(x_1 - y_1) f(x_2 - y_2) \geqslant f(x_1 - y_2) f(x_2 - y_1), \tag{A1.27}$$

that is,

$$f(z_2 - h) f(z_1) \geqslant f(z_1 - h) f(z_2), \tag{A1.28}$$

which proves (A1.25).

Consider now a function $f(x)$ that satisfies conditions (a) and (b) of the theorem. If it is identically zero, then it is a Pólya function. If not, it suffices to verify (A1.2) in the interval where $f(x) > 0$ (property (3)). Given x_1, x_2, y_1, y_2 such that $x_1 \leqslant x_2$ and $y_1 \leqslant y_2$ and that f is nonzero for the four arguments of (A1.2), we may define z_1, z_2, and h by (A1.29), with $z_1 \leqslant z_2$ and $h > 0$; (A1.28) thus entails (A1.31), and therefore (A1.30). Thus $f(x)$ is a Pólya function.

A.2 Totally Positive Functions of Order 2

A function of two variables $f(x, y)$ defined for $x \in \mathbf{X}$ and $y \in \mathbf{Y}$, where $\mathbf{X}$ and $\mathbf{Y}$ are totally ordered sets, is said to be totally positive of order k if and only if

$$\forall x_1, x_2, \ldots, x_i \in \mathbf{X}, \qquad x_1 \leqslant x_2 \leqslant \cdots \leqslant x_i,$$

$$\forall y_1, y_2, \ldots, y_i \in \mathbf{Y}, \qquad y_1 \leqslant y_2 \leqslant \cdots \leqslant y_i;$$

$$\begin{vmatrix} f(x_1, y_1) & f(x_1, y_2) & \ldots & f(x_1, y_i) \\ f(x_2, y_1) & f(x_2, y_2) & \ldots & f(x_2, y_i) \\ \cdots & \cdots & \cdots & \cdots \\ f(x_i, y_1) & f(x_i, y_2) & \ldots & f(x_i, y_i) \end{vmatrix} \geqslant 0. \tag{A2.1}$$

Generally, **X** and/or **Y** will be intervals of **R** or subsets of **N**, either finite or not.

We shall be particularly interested in totally positive functions of order 2, that is, those that satisfy

$$\text{(A2.2)} \qquad \begin{array}{ll} \forall x_1, x_2 \in \mathbf{X}, & x_1 \leqslant x_2, \\ \forall y_1, y_2 \in \mathbf{Y}, & y_1 \leqslant y_2 : \end{array} \qquad \begin{vmatrix} f(x_1, y_1) & f(x_1, y_2) \\ f(x_2, y_1) & f(x_2, y_2) \end{vmatrix} \geqslant 0 .$$

Totally positive functions of order 2 generalize the Pólya functions of order 2. Indeed, if $\varphi(x)$ is a Pólya function of order 2, the function

$$\text{(A2.3)} \qquad f(x, y) = \varphi(x - y)$$

is a totally positive function of order 2 on $\mathbf{R}^2$, of a particular type since it depends only on the difference between its two variables and since it is nonnegative. Conversely, if $\varphi(x - y)$ is nonnegative and totally positive of order 2 in $\mathbf{R}^2$, $\varphi(x)$ is a Pólya function.

Examples.

(1) The function

$$\text{(A2.4)} \qquad f(x, y) = (\lambda x - \mu y).\exp\left(\frac{-(\lambda x - \mu y)^2}{2}\right), \qquad \lambda > 0, \quad \mu > 0,$$

is totally positive of order 2 in $\mathbf{R}^2$.

(2) The function

$$\text{(A2.5)} \qquad f(x, y) = \frac{1}{x + y}$$

is totally positive of order 2 in $\mathbf{R}^2$.

A.3 Relation to IFR Functions

Theorem A.II. *The following four properties are equivalent:*

(a) $v(t)$ *is an IFR survival function;*
(b) $\Lambda(t)$ *(see (4.18)) is a convex function;*
(c) $v(t)$ *is a Pólya function of order 2;*
(d) $v(x - y)$ *is a totally positive function of order 2 in* $\mathbf{R}^2$.

We have seen in Section 10, Theorem 10.I, that (a) $\Leftrightarrow$ (b); and in Section A.2 that (c) $\Leftrightarrow$ (d). We show that (a) $\Leftrightarrow$ (c).

Let $v(t)$ be an IFR survival function. We have seen in Section 9 that then the failure rate by intervals defined by (9.10)

$$\mu(t\,;x) = \frac{v(t) - v(t + x)}{v(t)} = 1 - \frac{v(t + x)}{v(t)} \tag{A3.1}$$

is a nondecreasing function of t for all $x \geqslant 0$. It then follows that $v(t)/v(t + x)$ is also nondecreasing, and similarly that the function $v(t - x)/v(t)$ obtained by translation is also nondecreasing. According to Theorem A.I, $v(t)$ is then a Pólya function. The converse is proved in the same fashion.

BIBLIOGRAPHY

[1] Alven, W. H. von (ed.), *Reliability Engineering*. ARINC Research Corp., Prentice-Hall, Englewood Cliffs, New Jersey, 1964.

[2] Arrow, K. J., Karlin, S., and Scarf, H. (eds.), *Studies in Applied Probability and Management Science*. Stanford University Press, Palo Alto, California, 1962.

[3] Bansard, J. P., Le Glas, J. M., Maarek, G., and Valentin, F., Evaluation de la fiabilité d'un réseau de vannes. *Congrès National de Fiabilité*, Perros-Guirec, September 20–22, 1972. C.N.E.T., Paris, 1972.

[4] Barlow, R. E., and Marshall, A. W., Tables of bounds for distributions with monotone hazard rate. *J. Amer. Statist. Assoc.*, **60**(311), 872–890, 1965.

[5] Barlow, R. E., and Proschan, F., *Mathematical Theory of Reliability*. Wiley, New York, 1965.

[6] Bazovski, I., *Reliability Theory and Practice*. Prentice-Hall, Englewood Cliffs, New Jersey, 1961.

[7] Birnbaum, Z. W., and Esary, J. D., Some inequalities for reliability functions. *Proc. 5th Berkeley Symp. Math. Statist. Prob.* University of California Press, Berkeley, 1966.

[8] Birnbaum, Z. W., Esary, J. D., and Saunders, S. C., Multicomponent systems and structures and their reliability. *Technometrics*, **3**(1), 55–77, 1961.

[9] Birnbaum, Z. W., Esary, J. D., and Marshall, A. W., A stochastic characterization of wear-out for components and systems. *Ann. Math. Statist.*, **37**(4), 816–825, 1966.

[10] Buckland, W. R., *Statistical Assessment of the Life Characteristics*. Griffin's Statistical Monographs and Courses, No. 13, Hafner, New York, 1964.

[11] Calabro, S. R., *Reliability Principles and Practice*. McGraw-Hill, New York, 1962.

[12] Carlsson, S., and Grenander, U., Some properties of statistical reliability functions. *Ann. Math. Statist.*, **37**(4), 826–836, 1966.

[13] Chapouille, P., and De Pazzis, R., *Fiabilité des Systèmes*. Masson, Paris, 1968.

[14] Cozzolino, J. M., Probabilistic models of decreasing failure rate processes. *Naval Res. Logist. Quart.*, **15**(3), 361–374, 1968.

[15] Descamps, R., Prévision statistique des avaries et calcul des volants et rechanges. *DOCAERO*, **41**, 23–36, November 1956; **43**, 33–54, March 1957; **48**, 29–52, January 1958.

[16] Esary, J. D., and Marshall, A. W., System structure and the existence of a system life. *Technometrics*, **6**, 459–462, 1964.

[17] Esary, J. D., and Marshall, A. W., Coherent life functions. *SIAM J. Appl. Math.*, **18**(4), 810–814, 1970.

[18] Esary, J. D., and Proschan, F., Coherent structures of non-identical components. *Technometrics*, **5**, 191–209, 1963.

[19] Esary, J. D., and Proschan, F., Relationship between system failure rate and component failure rates. *Technometrics*, **5**, 183–189, 1963.

[20] Esary, J. D., Marshall, A. W., and Proschan, F., Determining an approximate constant failure rate for a system whose components have constant failure ratings. *In* D. Grouchko (ed.), Operations research and reliability. *Proc. NATO Conf., Turin, June 24–July 4, 1969*. Gordon and Breach, New York, 1971.

[21] Esary, J. D., Marshall, A. W., and Proschan, F., Some reliability applications of the Hazard transform. *SIAM J. Appl. Math.*, **18**(4), 849–860, 1970.

[22] Everett, H., III, Generalized Lagrange multiplier method for solving problems of optimum allocation of resources. *Operations Res.*, **11**(3), 399–417, 1963.

[23] Feller, W., *An Introduction to Probability Theory and Its Applications*, Vol. I, 2nd Ed. Wiley, New York, 1967.

[24] Gilbert, E. N. *J. Math. Phys.* **33**(1), 57–67, 1954.

[25] Grouchko, D. (ed.), Operations research and reliability. *Proc. NATO Conf., Turin, June 24–July 4, 1969*. Gordon and Breach, New York, 1971.

[26] Gumbel, E. J., *Statistics of Extremes*. Columbia University Press, New York, 1958.

[27] Haight, F. A., *Handbook of the Poisson Distribution*. Wiley, New York, 1967.

[28] Hansel, G., Contribution à la théorie de la sécurité de fonctionnement. Ph.D. thesis, University of Paris, 1961.

[29] Hansel, G., Nombre de lettres nécessaires pour écrire une fonction symétrique de *n* variables. *C. R. Acad Sci. Paris*, **261** (Groupe 1), 4297–4300, 1965.

[30] Hansel, G., Sur le nombre des fonctions booléennes monotones de *n* variables. *C. R. Acad. Sci. Paris, Ser. A*, **262**, 1088–1090, 1966.

[31] Hirsch, W. M., Cannibalization in multi-component systems and the theory of reliability. *In* J. J. Ferrier (ed.), *Large Scale Provisioning Systems*, English Universities Press, London, 1968.

[32] Jacobsen, S. E., and Arunkumar, S., Investment in series and parallel systems to maximize expected life. *Management Sci.*, **19**(9), 1023–1028, 1973.

[33] Kaufmann, A., *Cours Moderne de Calcul des Probabilitiés*. Albin-Michel, Paris, 1965.

[34] Kaufmann, A., and Cruon, R., *La Programmation Dynamique*. Dunod, Paris, 1965.

[35] Levy, P., Étude non paramétrique de la fiabilité des systèmes. Ph.D. thesis, University of Paris, 1973.

[36] Lloyd, D. K., and Lipow, M., *Reliability: Management, Methods, and Mathematics*. Prentice-Hall, Englewood Cliffs, New Jersey, 1962.

[37] Lomnicki, Z. A., Some remarks on the application of redundancy to system reliability. *European Meet. Statist., London*, 1966.

[38] Mann, N. R., A survey and comparison of models for determining confidence bounds on system reliability from subsystem data. *In* E. M. Scheuer (ed.), Reliability testing and reliability evaluation. *1972 NATO Conf. Proc.* California State University Press, 1972.

[39] Mine, H., Reliability of physical systems. *IEEE Trans. Information Theory*, **IT-S**, 138–150, 1959.

[40] Moore, E. F., and Shannon, C. E., Reliable circuits using less reliable relays. *J. Franklin Inst.*, **262**, 191–208 and 281–297, 1956.

[41] Morice, E., Quelque modèles mathématiques de durée de vie. *Rev. Statist. Appl.*, **14**(1), 45–126, 1966.

[42] Myers, R. H., Wong, K. L., and Gordy, H. M., *Reliability Engineering for Electronic Systems*. Wiley, New York, 1964.

[43] Neumann, J. von, Probabilistic logics and the synthesis of reliable organisms from unreliable components. California Institute of Technology, 1952; and *in* C. E. Shannon and J. McCarthy (eds.), *Automata studies* Ann. Math. Studies, No. 34, Princeton University Press, Princeton, New Jersey, 1956.

[44] Pollard, A., and Rivoire, C., *Fiabilité et Statistiques Prévisionnelles. La Méthode de Weibull*. Eyrolles, Paris, 1971.

[45] Price, H. W., Reliability of parallel electronic components. *IRE Trans. Reliability Quality Contr.*, pp. 35–39, 1960.

[46] Proschan, F., and Bray, T. A., Optimum redundancy under multiple constraints. *Operations Res.*, **13**(5), 800–814, 1965.

[47] Sandler, G. H., *System Reliability Engineering*. Prentice-Hall, Englewood Cliffs, New Jersey, 1963.

[48] Scheuer, E. M. (ed.), Reliability testing and reliability evaluation. *1972 NATO Conf. Proc.* California State University Press, 1972.

[49] Schwob, M., and Peyrache, G., *Traité de Fiabilité*. Masson, Paris, 1969.

[50] Shooman, M. L., *Probabilistic Reliability. An Engineering Approach*. McGraw-Hill, New York, 1968.

[51] Simon, R. M., Optimal cannibalization policies for multicomponent systems. *SIAM J. Appl. Math.*, **19**(4), 700–711, 1970.

[52] Simon, R. M., The reliability of multi-component systems subject to cannibalization. *Naval Res. Logist. Quart.*, **19**(1), 1–14, 1972.

[53] Smith, W. L., Renewal theory and its ramifications. *J. Roy. Statist. Soc., Ser. B*, **2**, 243, 1958.

[54] Srinivasan, V. S., The effect of standby redundancy in system's failure with repair maintenance. *Operations Res.*, **14**(6), 1024–1036, 1966.

[55] Weibull, W., *Ing. Wetevskaps Akad. Handl.*, **151**, 1938.

[56] Weibull, W., *J. Appl. Math.*, **18**, 293–297, 1951.

[57] Weiss, G. H., A survey of some mathematical models in the theory of reliability. *In* M. Zelen (ed.), *Statistical Theory of Reliability*. University of Wisconsin Press, Madison, Wisconsin, 1963.

[58] Wilcox, R. H., and Mann, W. C. (eds.), *Redundancy Techniques for Computing Systems*. Spartan Books, New York, 1962.

[59] Zelen, M. (ed.), *Statistical Theory of Reliability*. University of Wisconsin Press, Madison, Wisconsin, 1963.

INDEX

Numerals in boldface indicate section numbers. Lightface numerals give page numbers; those in italics refer to pages with definitions.

A
B 7
C 8
D 9
E 0
F 1
G 2
H 3
I 4
J 5